In the Footsteps of Programming Teachers

Krzysztof Frankowski

To my students, who listened to my lectures and asked many intelligent questions about the beginnings of Computer Science.

IN THE FOOTSTEPS OF PROGRAMMING TEACHERS.

This work is distributed under a Creative Commons license, Attribution-NonCommercial-NoDerivatives 4.0 International (CC BY-NC-ND 4.0). This means you are free to share this work for non-commercial purposes with no changes (derivatives), as long as you give appropriate attribution. For the detailed terms, see https://creativecommons.org/licenses/by-nc-nd/4.0/.

Edited by Dan Frankowski.

Cover illustration by Anetta Dobrakowska.

Cover production by Susan Everson.

ISBN 978-1-105-46202-3.

Contents

	Introduction	v
1	**In the footsteps of John McCarthy** *the MIT professor who invented Lisp, the first functional programming language*	1
2	**In the footsteps of Heron of Alexandria** *who in ancient times had already described a very accurate method of computing square roots*	7
3	**In the footsteps of John von Neumann** *who first described in detail how to construct a modern computer*	13
4	**In the footsteps of George Boole** *who in 1854 described the elements of logic in his book "The Laws of Thought"*	21
5	**In the footsteps of Donald Knuth** *who taught us mathematical analysis of programs*	31
6	**In the footsteps of John Backus** *who in 1954 working for IBM invented FORTRAN, the oldest higher level programming language still used today*	41
7	**In the footsteps of Isaac Newton** *who taught us how to raise a sum to the n-th power*	51

8 In the footsteps of James Wilkinson
who taught us analysis of numerical errors   57

9 In the footsteps of Brook Taylor
who in 1715 taught us how to study and calculate a power series   65

10 In the footsteps of Archimedes
who in ancient times computed the number π accurately   71

11 In the footsteps of Euclid
who over 2000 years ago knew the beauty of proof and of number theory   81

12 In the footsteps of Euler
who in 1759 published an analysis of the knight's tour problem   97

13 In the footsteps of Keith Clark and Robert Kowalski
inventors of Prolog, a language used in rule-based programming   111

Appendices   121

A In the footsteps of John Backus and Peter Naur
who invented a notation for describing the syntax of programming languages   121

B In the footsteps of Gottfried Leibniz
who in the 17th century taught us the binary representation of numbers   125

C In the footsteps of Steven Wolfram
inventor of Mathematica, the language used in this book   131

D In the footsteps of the Babylonians
who around 1800 BC showed us how to solve quadratic equations   139

E Online judges   145

F Mathics   147

Introduction

This book is based on lectures I gave at the University of Łódź in Poland. I called them *Master Programming*. They were intended for first-year students and others who wanted to improve their programming skills. The only prerequisites were Calculus 1 and knowledge of programming in a C-like language. My purpose was to teach students how to write correctly-working small programs and to encourage them to send their solutions to international competitions judged by robots, since students not experienced in programming do not debug their programs thoroughly.

The lectures were also intended to help listeners become masters. A master is someone who knows how to do something better than other people. A necessary condition for mastery is a fondness for the craft being mastered and an intuition about what constitutes beautiful work, in this case, elegant programs.

We live in times of awful extravagance and waste of resources. Indeed, our actions and attitudes may be threatening even the beauty and human habitability of the natural environment of our planet. Mindset and values are important from small to large. So I think that it is important to teach young people to be parsimonious; that is, to use only these resources that they need to accomplish a given task. I want to teach students to recognize the beauty of small pearls rather than tolerate the wasteful tangles that result from laziness. I am not interested in saving one microsecond here or there. I am promoting an approach that requires thinking about whether we can do some task better with fewer resources. In this goal my inspiration was always Professor Donald Knuth whose every program is a polished pearl.

Without formal proofs we cannot be sure that our programs deliver correct results for every possible input. Additionally, it is easier to hide errors in badly written, ugly programs than in elegant parsimonious ones.

I call this book "In the Footsteps of Programming Teachers" to emphasize the influence of scientists, ancient and modern, on my way of thinking and also to spare my readers and myself the burden of many references to the computing literature. I do not claim that the ideas I present here are original. In the era of the Internet it is enough to search a few minutes to find information that, at the beginning of computer science, was passed on by word of mouth and, in most cases, had many fathers.

Here is an overview of the topics in each chapter.

Chapter One — *In the footsteps of John McCarthy* is an introduction to functional programming. It begins with an informal introduction to functional programming or to using the concept of multivariate functions, the pseudo-function of choice 'If', and elementary arithmetic operations. I depart from the custom of using prefix notation to represent these operations, as Lisp does, because I believe this is an artificial formulation for students. I use the replacement statement to indicate 'remembering' the value of a function. I use the computer language *Mathematica* for my example programs, thus I have no trouble printing results since *Mathematica* is easily instructed to print them.

Chapter Two — *In the footsteps of Heron of Alexandria* is a more complicated example of functional programming. It contains the application of different functions to the computation of the square root using Heron's method. I study Heron's reasoning and imitate it, step by step, in an example program.

Chapter Three — *In the footsteps of John von Neumann* is an introduction to the imperative programming of a binary machine. It introduces the concept of **machine memory** in which programs as well as data are represented in binary form and imprisoned in cells, called words, of a constant size or word length. It also introduces the concept of number **types** and their representations, and programming actions such as a change of the state of a machine's memory. A change to the state of the machine's memory is achieved by **replacement** which I describe in detail. The 'For' instruction is used in order to repeat a series of operations many times; that is, to execute a program **loop**. Examples are repeated in **functional** and **imperative** styles to illustrate their differences.

Since programs are becoming more complicated we need a method of checking that they produce correct results at least in the cases checked, so I introduce the concept of a **trace**, hand-checking the program by computing some result for several iterations of a loop to trace the series of values that the loop produces.

Chapter Four — *In the footsteps of George Boole* is about the elements of propositional calculus. It presents those elements of propositional calculus needed to construct correct 'If' and 'For' instructions. The chapter describes four operations: negation, equivalence, disjunction and conjunction. It also presents conditional disjunction and conditional conjunction and their use. The chapter also explores the use of propositional calculus to validate that the arguments to a function being computed are in the proper range.

Chapter Five — *In the footsteps of Donald Knuth* is an introduction to the analysis of the time complexity of programs. The complexity of functional programs is measured by the number of calls to the program's main function. The complexity of imperative programs is measured by the number of repetitions of the program's main loop. Examples show how to reduce the number of repetitions of a loop and how to achieve logarithmic complexity in computing both the integer power of a number and Fibonacci numbers.

Chapter Six —*In the footsteps of John Backus* is about the use of arrays of numbers. It covers these ideas: saving the results of a computation and ordering them in a sequence; local, global, automatic and static variables; and the use of arrays in **lazy** and **dynamic** computations. Examples include linear and binary search.

Chapter Seven — *In the footsteps of Isaac Newton* is about the computation of binomial numbers. This computation is used to show how to estimate the size of an integer result, the danger of large integers in programs and **integer overflow**, sometimes called **overspill**. I show the computation of binomial numbers using factorials, and a loop of multiplications and divisions; how to use Stirling's formula to estimate the size of any factorial and of the largest binomial number representable in a given machine; and the application of dynamic computation to computing binomial numbers using the Pascal triangle, and the use of symmetry to reduce the size of an array of binomial numbers.

Chapter Eight — *In the footsteps of James Wilkinson* is an introduction to the analysis of floating point error. I examine the world of floating point numbers: floating point numbers as a poor substitute for mathematical real numbers; absolute and relative error; the failure of floating point numbers to satisfy arithmetic laws such as distribution and associativity; the IEEE representation of floating point numbers; a simplified model of error analysis; and estimates of the error of several floating-point calculations.

Chapter Nine — *In the footsteps of Brook Taylor* is about the computation of the power series. Topics include a convenient method of computing the coefficients of the Taylor series expansion; a program to compute exponentials $exp(x)$; remarks about determining when to finish the calculation of an infinite series or computation of limits; computation of the indeterminate values $0/0$ and ∞/∞, and the difference of large numbers that are close together— that is, $\infty - \infty$; and Horner's formula for computing polynomials.

Chapter Ten — *In the footsteps of Archimedes* is about the use of error analysis in the computation of π. This chapter covers several methods for computing π and related topics: the classical computation of π using regular polygons inscribed in and circumscribed around a circle; generalizations of the Archimedean method by Snellius, Pfaff, and using arithmetic and geometric means; computations of the area and circumference of a circle using only the four basic arithmetic operations and extraction of square root; a 'proof' that $\pi = 0$; and the history of different methods of π computation by A. Sharp, I. Newton, J. Machin, L. Euler, D. Bailey, and T. Kanada.

Chapter Eleven — *In the footsteps of Euclid* is about number theory programs. Number theory applications include: various methods of computing the largest common divisor of two natural numbers using the Euclidean algorithm; a generalization of the Euclidean algorithm; the computation of prime numbers using the Sieve of Eratosthenes and by the search for divisors; checking whether a number is prime; computing $x^n \bmod m$ quickly for natural numbers; a few remarks about Fermat's theorem; pseudo-prime numbers; quick computation of Perrin numbers; and the application of prime numbers to some RSA code encryptions.

Chapter Twelve — *In the footsteps of Leonhard Euler* is about programming backtracking algorithms. The chapter illustrates backtracking by finding a path through a maze; a search for all solutions to the example problem of placing eight queens on a chessboard; a search for one of many solutions to a backtracking problem; several attempts to solve the problem of knight's covering the entire chessboard once; illustrating, by several attempts to speed up the knight's tour problem, the impossibility of substantially reducing the exponential time complexity of backtracking algorithms; and, success in solving the knight's problem searching for narrow passages using the analysis of a static table of all the possible knight's moves.

Chapter Thirteen —*In the footsteps of Keith Clark and Robert Kowalski* is about the **rule-based** programming paradigm. One can write programs by giving the computer a set of rules (formulas) and allowing the computer to choose which rule to apply based on the form of the inputs. The chapter covers writing named and unnamed functions (rules) with a variable number of parameters, with different types of parameters, and with unnamed parameters; and conditional calling of functions. The examples include differentiation of polynomials; use of functions such as 'Map' and 'Apply'; rule-based programs for summation, sorting and other programs first introduced in previous chapters.

The book also contains six appendices. **Appendix A** explains the Backus Normal Form notation for the syntax of context free grammars. **Appendix B** describes how numbers are represented in computers. **Appendix C** gives details of the computer language *Mathematica* and an index of the functions used in this book. **Appendix D** discusses the programming of a quadratic equation solver. **Appendix E** briefly describes the world of online programming judges. **Appendix F** quickly shows Mathics, a free, open-source alternative to Mathematica.

Finally, a few words about the example programs in this book: all are written in *Mathematica* because that language supports easy expression of many programming paradigms. I highly recommend *Mathematica* to all engineers for approximate and symbolic computations and for other numerical applications. *Mathematica* is the slide rule of the 21st century.

In spite of the fact that I use and like *Mathematica*, it is a commercial product and not generally available (although a free and open-source program Mathics uses the same syntax and has many identical functions). I demand from my students only one working program in that language to be sure that they can read the examples in this book. They may write their programs in any language they choose, for example C, C++, or Java.

When choosing topics I was guided by the attempt to answer the question of what should be taught to fledging, intelligent student. It seems to me that since the structure of the data in small programs is usually simple, we can emphasize their design. Students should know:

- The three programming paradigms: functional, imperative and rule based

- Basic data types: integers, boolean, reals and their differences in the two worlds of mathematics and computer science

- The concepts of program complexity and order of magnitude

- Organization of data in one- and two-dimensional arrays

The first two chapters of the book describe functional programming with integers and reals; the third, imperative programming; the fourth discusses the laws of logic. The fifth chapter deals with time complexity; the sixth chapter, the use of arrays. The seventh chapter is devoted to the problem of integer overflow. The next three chapters discuss the programs with real numbers and computation of limits. I devote the last chapter to programming using rules. Chapter eleven was included due to the beauty of number theory, as an opportunity to present probabilistic algorithms, and to apply modular arithmetic to the theory of RSA codes. Chapter twelve (backtracking programs) describes shortening of program execution times by adding some tips to straightforward recursion.

I assume that my students have already written programs, so the topics often discussed in books about programming can be treated superficially, but subjects often omitted I expand further.

In normal classes (two hours of lecture plus two hours of lab per week in one semester) chapters eleven and twelve were omitted, but in a seminar-style class material in these chapters was very popular.

If I had to pick one thing for the audience of my lectures to remember, it would be the great significance of validation: i.e., checking that the arguments of a function $f[x_1, x_2, \cdots, x_n]$ used in a program satisfy the specifications of the problem being solved. Programmers are engineers and as a bridge builder is responsible for its sound design, developers are responsible for the results of programs they write.

Chapter 1

In the footsteps of John McCarthy,
the MIT professor who invented Lisp, the first functional programming language

(Elements of functional programming)

Programming is a method by which we communicate with computers to force them to do things that we call computations. We may also say that programs steer computations.

You are correct when you think that we start this book on weak foundations: the terms 'computation' and 'steer' are not defined.

Imagine that a computation is a magic creature that lives in a computer and uses data, which also reside in the computer, as food. Programs are shepherds that command computations to consume data.

In this magic kingdom of computers data may be eaten many times, in its original state or transformed by the computation. We call the commanding of the computation to transform data 'steering.'

Programs are written by people in order to steer computations. When we write a program that correctly steers a computation, we will get a desired result after multiple transformations of the data by the program. In order for people to write programs, they need to communicate with computers in a language they (the people) understand. A machine that performs computations understands very few orders, and the language it understands is uncomfortable for people to use. That language is composed of a sequence of zeros and ones. The machine's great advantage is its extraordinary speed and accuracy in performing computations.

Because people usually write programs, programming languages are adapted to the way people think instead of the way machines operate. You are correct when you believe there is not a big difference between writing programs and casting magic spells. I shall try to teach you to cast spells correctly, and not to repeat Goethe's story about the sorcerer's apprentice who forgot the proper spell for making the broom bring water. He nearly flooded the whole house.

A paradigm is the way to think about the kinds of orders to use to force the machine to carry out computations. The functional paradigm is the simplest computational paradigm. It uses the concept of function to transform inputs into desired results.

In order for our spells to express our intention, they have to be simple and unambiguous. We cannot expect to master complicated spells from the beginning, so we employ mathematics to help us construct a few simple spells.

The mathematical help that leads to a few simple spells is the concept of a function. We assume that the machine knows what a function is.

We write our programs using multivariate functions of the form $y = f_1[x_1, x_2, ..., x_n]$, where we call the sequence $x_1, x_2, ..., x_n$ parameters, input values, or inputs, and we call y the result of applying the function to the parameters, also called the function's value. (Although McCarthy invented Lisp, we are going to use the notation of Mathematica.) The result of the computation, y, may now be input to another function, for example $y_2 = f_2[y, x_1, x_2, ..., x_n]$. Obviously

y_2 may be input to another function, and so on.

The result of a computation is obtained by applying the sequence of functions $f_1, f_2, ..., f_k$ in such a way that the application of the final function f_k yields the desired result.

The most common functions used here are the four basic arithmetic operations: addition, subtraction, multiplication and division. We call them the fundamental operations, and we assume that the machine knows them. For simplicity we write these four operations using infix notation. Hence we write $y = a + b$ instead of $y = +[a, b]$ as we would if we use the notation $y = f_1[x_1, x_2, ..., x_n]$. We might also write $y = Plus[a, b]$. We assume that the computer knows the priority of the fundamental arithmetic operations. Therefore $y = 4 + 2 * 3$ yields 10 (first multiply 2 and 3, then add 4) and not 18, the result we would get if we apply the operations from left to right ignoring priorities. We use parentheses to change the order of applying operations. For example, $y = (4 + 2) * 3$ would indeed yield 18.

Note that we use parentheses only for grouping expressions and never in a function, so $f(x)$ always means $f * x$ and never $f[x]$.

After a call to the function $f[e_1, e_2, \cdots, e_n]$, the computation of $f[x_1, x_2, \cdots, x_n]$ starts with substitution of the values of expressions $e_1, e_2, \cdots, e_n$ (called arguments) into the corresponding parameters $x_1, x_2, \cdots, x_n$.

Some functions change this order of evaluating parameters or execute additional tasks, therefore they have only the form of a function. We will call them pseudo-functions or **key words** and use them as spells. The choice function 'If' is such a spell. It has the form:
If [<condition> , <action for True> , <action for False>].

'If' always evaluates the first argument, <condition>, and then computes either <action for True> or <action for False> depending on the result of <condition>.

Words placed between '<' and '>', for example <condition> or <action for False>, are called meta-names, and are used to denote the types of expressions that may occur in that position in the list of arguments to the function.

Here is an example of using 'If' to compute the absolute value of x:

$$abs[x\_] := \text{If}[x < 0, -x, x].$$

This function is defined for arbitrary x (note the use of '_' after the name of the parameter to indicate an arbitrary argument to *abs*), and ':=' means 'is defined as.' Writing $y = abs[-5]$ means that y is the result of applying *abs* to the argument '−5.'

Here is an example of a functional program that computes Fibonacci numbers, and a call to that program. Fibonacci numbers are defined by the recurrence relation

$$f_1 = f_2 = 1, \quad f_k = f_{k-1} + f_{k-2}, \quad \text{for } k = 3, 4, \ldots, n. \quad (1.1)$$

Here are the first ten Fibonacci numbers:

$$1, 1, 2, 3, 5, 8, 13, 21, 34, \text{ and } 55.$$

Illustration 1.1 *Functional Fibonacci.*

```
fibonacci[n_]:=
  (* Compute the n-th Fibonacci number fn *)
  If[n < 3,
     1,
     fibonacci[ n-1] + fibonacci[ n-2] ];

fibonacci[6]    8
```

Note that a call to `fibonacci[6]` yields the result 8.

Because this book uses only a few selected phrases of *Mathematica*'s rich vocabulary, we can easily summarize the rules for using them:

1. Names of functions and variables begin with a letter, and may be followed by an arbitrary number of letters or digits. (Names of objects, functions or constants, in the *Mathematica* library always start with a capital letter.)

2. We write the four basic arithmetic operations (addition, subtraction, multiplication, and division) in the common infix notation instead of prefix, or functional, notation. The symbol for the operation occurs between the operands. The priority of these operations is defined by the normal mathematical rules. We use the functions 'Round' or 'Quotient' to get integer results from division.

    The result of integer division using *Mathematica*'s normal division function is usually a fraction.

3. We indicate a sequence of computations of functions by separating them with semicolons. For example, $y1 = f1[...]; y2 = f2[...]$; ...

4. A function is defined by giving its name, followed by a list of names (the function's parameters) placed between brackets. Each parameter ends in an underline character. Next we write ':=' (defined by). A function is called with arithmetical expressions (called arguments) in place of the parameter names.

5. 'Choice' is indicated by the function
If[<condition> , <action for True> , <action for False>].

   The <condition> is computed using the relations '<' (less than), '>' (greater than) '==' (equal), '!=' (not equal), '<=' (less than or equal to) and '>=' (greater than or equal to). Evaluating the condition yields the values True or False.

6. Comments are written (* <comment string> *) and do not influence the computation.

As we see the only word used to cast a spell is the word 'If.' Its spell is expressed by:
If [<condition> , <action for True> , <action for False>].
The additional syntax forms are

- the brackets '[' and ']' which denote the beginning and end of the list of parameters or arguments to that function

- the underline '_' character terminates every parameter, denoting arbitrariness

- the semicolon ';' separates sequences of called functions

- the ':=' defines a function

- and '(*' and '*)' are used for comments

Appendix A gives a more detailed description of these rules using BNF (Backus Normal Form).

Chapter 2

In the footsteps of Heron of Alexandria,

who in ancient times had already described a very accurate method of computing square roots

(A more complex functional program)

We are going to write a more complex functional program to compute the square root of a real non-negative number using Heron's formula.

Even before we start our programming task, we have a difficulty in trying to define a square root. This difficulty shows the difference between mathematicians and programmers. The definition "$y = \sqrt{x}$ is such a number $y \geq 0$, that $y^2 = x$" is a reasonable definition for a mathematician. It is short and accurate. For a programmer it is nearly useless. It does not say anything about what to do with the number x to compute its square root. Even if we could find such a y, how are we going to represent it to use in further computations?

Finally when we decide to compute $\sqrt{x}$ with a given accuracy, do we have a right to call it a square root? The problem of finding 10 decimal figures of a square root is a completely different problem than finding 1000 figures.

There are many differences between the mathematician's and the programmer's points of view. We mentioned two:

1. Defining a function by its inverse, which assures its existence, but says nothing about computing it.

2. Infinite accuracy versus finite representation.

There are more: efficiency in representation and efficiency in computation.

Mathematicians are like gods. They say, "Let A be a matrix!" and at that moment they create a matrix. They can talk about it until the end of the lecture.

Programmers are engineers. Even before they start writing programs about matrices they ask many questions: How big can this matrix be? What kind of elements will it have? What operations do we need? Is it dense, or maybe it has lot of zeros? Then, after they start to outline their program, they still worry about whether the inputs satisfy the proper conditions of the problem at hand. They cannot believe in the truth of anything.

Ancient Babylonians used square roots thousands of years ago in geodesy and other practical land measurements. They constructed extensive tables of squares of integers, which they used for square root approximations. Heron of Alexandria was the first scientist to describe a method of computing a square root. That method is used to this very day.

He writes: *"Since 720 is not a full square, we can obtain its side with a small error in the following way: the next full square is 729, who's side is 27. Divide 720 by 27. This gives $26\frac{2}{3}$. Add to it 27, obtaining $53\frac{2}{3}$ and take the half, obtaining $26\frac{5}{6}$. If we multiply $26\frac{5}{6}$ by itself we obtain $720\frac{1}{36}$, so the difference in the square is $\frac{1}{36}$."*

Reading Heron's reasoning about square root approximation and repeating his method using a sequence of better and better approximations, we hope to get a square root as accurate as we need. We formulate the following plan to compute $\sqrt{x}$:

1. Build the sequence of approximations s_n starting with some close initial value $s_0 = a$.

2. Then compute the sequence

$$s_{n+1} = 0.5\left(s_n + x/s_n\right) \text{ for } n = 0, 1, \ldots \tag{2.1}$$

We use this sequence to write a functional program. As a stopping rule, we check (as Heron did) if the error in the square is small enough.

Illustration 2.1 *Square root according to Heron.*

```
sqrtFun [x_, approx_] :=
  (* Compute square root using Heron's method *)
  If[good[x, approx], approx,
    sqrtFun[x, better[x, approx]]
  ];
better[x_, approx_] :=
  (* Approximation is improved by averaging *)
  0.5 * (approx + x / approx);

good[x_, approx_] :=
  (* Check if the approximation is good enough *)
  (abs[approx * approx - x] < 0.000001);

abs[x_] := If[x < 0, - x, x];

sqrtFun[2,1]    1.41421     .
```

This program follows Heron's reasoning very closely and gives an answer 1.41421 for $\sqrt{2}$. How good is this answer? We are interested in the number of accurate digits. For instance, in a number $x = 1.4153$ the first three digits 1, 4, 1 are accurate digits of $\sqrt{2} \approx 1.41421352\ldots$ To find the approximate number of accurate digits, Ad, we first compute the relative error

$$R(x) = (x - x^*)/x^*$$

where x is the computed number and x^* is the exact number. In our example

$$R(x) = (1.4153 - 1.41421)/1.41421 \approx 0.00077 = 0.77 \; 10^{-3}.$$

Next compute

$$Ad = -\log_{10}(|R(x)|).$$

In our case $Ad = 3.11...$, so we have three accurate digits. In modern times we do not have as much faith in our intuition as Heron did, so we should ask many questions:

1. Does the sequence s_n given in (2.1) converge?
2. If yes, how quickly does it converge?
3. Does it converge to $\sqrt{x}$?
4. How accurate is the final result?
5. How can we change the condition $|approx^2 - x| < \epsilon$ (error in square) to $|s_n - \sqrt{x}|/\sqrt{x} < \eta$ (relative error in $\sqrt{x}$)?
6. How do we find a first $approx$?

To answer these questions, let us define the relative error of s_n (for $x > 0$) as

$$R_n = \frac{s_n - \sqrt{x}}{\sqrt{x}} \qquad (2.2)$$

Substitute $s_n = \sqrt{x}(1 + R_n)$ into equation (2.1) to obtain, after some algebra, the formula for R_n

$$R_{n+1} = 0.5 \left(\frac{R_n^2}{1 + R_n} \right) \quad \text{for } n = 0, 1, ... \qquad (2.3)$$

We start with

$$s_0 = a > 0, \; R_0 = \frac{a - \sqrt{x}}{\sqrt{x}}, \; R_1 = \frac{0.5 R_0^2}{1 + R_0} = 0.5 R_0^2 \cdot \sqrt{x} > 0 \qquad (2.4)$$

therefore $R_n > 0$ for $n > 0$.

Consider two cases:

1. $R_1 < 1$. Then

$$R_n < R_{n-1}^2/2 < (R_{n-2}/2)^2 /2 = R_{n-2}^4/8 \qquad (2.5)$$

The number of correct digits more than doubles with each step. To show that, assume $R_n = 10^{-k}$. Then $R_{n+1} \approx 0.5 \times 10^{-2k}$.

2. $R_1 \geq 1$. Then
$$R_{n+1} = \frac{1}{2}\frac{R_n}{1+R_n^{-1}} < 0.5 R_n \qquad (2.6)$$

In this case the error R_n is smaller than half of the previous error. Therefore, after a few steps there has to be such an N that $R_N < 1$, and we are again in the first case.

Since the sequence $R_n = (s_n - \sqrt{x})/\sqrt{x} \to 0$ for $n \to \infty$, s_n converges to $\sqrt{x}$.

Therefore we have proved

Theorem 2.1
Heron's sequence s_n, given by $s_{n+1} = \frac{1}{2}(s_n + x/s_n)$, converges for every starting $s_0 = a > 0$ to $\sqrt{x}$, and if our approximation s_0 is sufficiently good, then the number of significant figures doubles with each iteration.

Now let us deal with the fifth question, namely how to change Heron's termination condition $|approx^2 - x| < \epsilon$ into $|s_n - \sqrt{x}|/\sqrt{x} < \eta$, the relative error of s_n, even if we do not know $\sqrt{x}$.

Let $\epsilon = approx^2 - x$. Therefore $approx = \sqrt{x+\epsilon}$ and we can deduce

$$\frac{\sqrt{x+\epsilon} - \sqrt{x}}{\sqrt{x}} = \frac{\sqrt{x}(\sqrt{1+\epsilon/x}) - \sqrt{x}}{\sqrt{x}} = \sqrt{1+\epsilon/x} - 1 \approx \frac{\epsilon}{2x} < \eta \qquad (2.7)$$

Here we used the formula $\sqrt{1+z} \approx 1 + z/2$ for small $|z|$. If we want to have at least 5 significant figures the function *good* should be replaced by function

```
good[x_, approx_]:=
    2 * x * abs[approx * approx - x] < 0.000005
```

If we start with $a > \sqrt{x}$ (as Heron did), we can skip the $abs[\cdots]$ function, because the sequence s_n converges from above.

Finally we come to the last question, how to choose the first approximation a. The answer to this question depends on the representation of the number x itself. It is useful to separate out digits of accuracy in representation from the magnitude of the number itself. In this way, we will be able to represent large and small numbers with the same accuracy.

Suppose that $x = b * 2^n$, where b ($0.5 \leq b < 1$). We will call b the *mantissa* of the number x, and n its *exponent*. For the majority of modern computers this is very close to the true representation, as we explain in Appendix B. Then $y = \sqrt{x} = \sqrt{b} * 2^{(n/2)}$. The exponent of y will be $n/2$, or $(n+1)/2$, depending on whether n is even or odd, and our problem of finding an a is reduced to the approximation of $\sqrt{b}$ for $0.25 \leq b < 1$.

To have 5 significant figures after 3 iterations, we have to have an initial guess good to 0.7 significant figures. If we take, for instance, $a = 0.6$ for $0.25 \leq x < 0.5$ and $a = 0.9$ for $0.5 \leq x < 1$, we can guarantee 5 significant digits after 3 iterations.

We can only speculate on how Heron obtained formula (2.1). Heron calls his formula the Babylonian method, and uses only one iteration of (2.1) to compute $\sqrt{x}$ for x a natural number. Babylonians constructed a large table of squares of natural numbers, which they used even for multiplication, using the formula

$$a \times b = \left(\frac{a+b}{2}\right)^2 - \left(\frac{a-b}{2}\right)^2. \qquad (2.8)$$

For instance, $7 \times 5 = 6^2 - 1^2 = 35$.

Substitute in (2.8) $a = s_0$ and $b = x/s_0$ to obtain

$$x = 0.25(s_0 + x/s_0)^2 - R, \qquad (2.9)$$

where $R = (s_0 - x/s_0)^2$ is a small error that one can omit.

Chapter 3

In the footsteps of John von Neumann,
who first described in detail how to construct a modern computer

(Elements of imperative programming)

Imperative programming is less abstract and closer to the machine's language than functional programming. Here we assume the machine model known as the *von Neumann machine*, named for the Hungarian mathematician John von Neumann. As a refugee living in USA, he worked at the Institute of Advanced Studies in Princeton, New Jersey, and, in 1942, for the first time, described a modern computer in detail.

In his machine *programs* and *data* live in *memory*, and are represented digitally using a *binary* number system. In this system every natural number N is written as

$$N = a_k 2^k + a_{k-1} 2^{k-1} + \cdots + a_1 2 + a_0 \qquad (3.1)$$

In this notation N has $k+1$ binary digits, where each a_i is either zero or one, with the exception of a_k, which is always one. For example, the number $10 = 2^3 + 2$ is represented as 1010_2 using four binary digits also called **bits**. More information about this representation, and some operations on binary numbers, is given in Appendix B.

Our present computers have much more memory and are many orders of magnitude faster than von Neumann's machine, but are surprisingly similar. They consist of a control unit, an arithmetical unit, and memory which is addressed sequentially as $m_1, m_2, \cdots, m_n$, where the values m_i can be changed by the executing program. The control statements which the machine executes can change values in the part of memory m_i that holds *data*.

We still use functions in our computations, but we add two more concepts: **replacement** (overwriting the data in a memory cell) and a **loop** (repetition of a sequence of functions). In this style, called *imperative*, the process of computation is different than the functional style. Imagine the replacement instruction as the change of the state of the memory cell m_i. Up to now '=' was used occasionally as a convenience for remembering an often used partial result in $y = f[...]$, such as in

Illustration 3.1 *Area of triangle.*

```
tArea[a_, b_, c_]:=
  (* Given 3 sides of triangle, compute its area. *)
  (p = 0.5(a + b + c); Sqrt[p(p - a)(p - b)(p - c)]);

tArea[3,4,5]  6.0
```

Now the same '=' denotes a change of the value in the cell named y (in our example p).

Such a y occupies a fixed number of bits determined by its **type**. Therefore integers, which were up to now arbitrarily large, have to be defined more accurately. Following the C language, we define a **short** to be 8 bits, an **int** to be 32 bits, and a **long** to be 64 bits.

Therefore the numbers

$$MaxShort = 2^7 - 1 = 127,$$
$$MaxInt = 2^{31} - 1 = 2,147,483,647, \qquad (3.2)$$
$$MaxLong = 2^{63} - 1 = 9,223,372,036,854,775,807$$

are of fundamental importance in programming, because they define the limits of validity of algebraic laws for integers used in the computation. It can happen that $MaxInt + 1$ becomes a negative number, giving the wrong results when it is used in subsequent computations. This phenomenon is called *integer overflow*, or sometimes *overspill*.

Negative integers are usually represented as the so-called *two's complement* of the corresponding natural numbers. For instance for number 10 represented as short by 00001010_2, -10 becomes 11110110_2. Details of this representation are given in Appendix B.

Responsibility for the use of such numbers rests with programmers. Many millions of dollars can be lost by careless programs, as witnessed by the fall of the European satellite Ariane 5, caused by an overspill.

Previously the names of arguments of the function $y = f(\vec{x})$ (where the vector $\vec{x}$ is a list of parameters) had to be different from the variable's name y. A function could not change the value of y, only compute it. In imperative programming this restriction is removed, so we may have $w = f[w, \vec{x}]$.

We can describe the action of $w = f[w, \vec{x}]$ as a pair of instructions: $w_i = f(w_{i-1}, \vec{x}); w_{i-1} = w_i$. The second instruction makes it impossible to use the previous w, which disappears forever.

The expression $x = x + 1$, which in functional programming makes no sense, in imperative programming means that the value of the variable x, which is in memory, is incremented by one, and the previous value is destroyed. Incrementing by one is used so often that many languages have a special notation for this operation. Following a convention of the C language, we will write $x++$ instead of $x = x + 1$. In a similar way, we write $+=$, $-=$, $*=$ and $/=$ as abbreviations. For instance, we write $x += 2$ in place of $x = x + 2$. We also write $x--$ instead of $x = x - 1$.

In imperative programs each variable is a triple (name, value, type), and a computation is the process by which we change the value of a given pair (name, type). We accomplish this using *replacement*, which often occurs in a *loop*.

The concept of *loop* is fundamental to the imperative paradigm, and has its own spells. We will use the most general form:
For [<start>,<condition>,<change of variables>,< orders>].
For executes <start> first and then repeats execution of the pair: (<orders>, <change of variables>) as long as the <condition> is True. When the <condition> is False, the loop stops executing. If the <condition> is False initially, the loop never executes.

Example:
Compute the sum

$$s = \sum_{i=1}^{n} a_i = a_1 + a_2 + \cdots + a_n. \tag{3.3}$$

We can use the recurrence relation

$$s_0 = 0; \quad s_i = s_{i-1} + a_i, \quad \text{for} \quad i = 1, 2, \ldots, n. \tag{3.4}$$

For the specific example we will set $a_i = i^2$ and compute

$$ss = \sum_{i=1}^{n} i^2$$

The following imperative program shows this computation:

Illustration 3.2 *Sum of squares.*

```
ss[n_Integer] :=
  (* Compute ss = sum i*i, for i = 1, 2, ... n *)
  Module[{sum = 0, i},
         For[i = 1, i <= n, i++, sum += i * i];
         sum];

ss[3]  14
```

The order **Module[{x, y, ...}, <expr>]** is also a spell that assures us that the names x and y (which can be initialized using replacement) in <expr> are local, i.e., are used only inside this function, and do not clash with the same names used outside. Locality of names, also called *scope* or *name scope*, is very important. It allows us to write a program to compute a function once and use it in many programs without worrying that the same names will be spoiled in other programs. We added the type of parameter _Integer to limit the domain of the function ss from arbitrary type to type Integer only.

We can compute the sum of squares using the formula

$$\sum_{i=1}^{n} i^2 = \frac{n(n+1)(2n+1)}{6}$$

that we can prove by mathematical induction, but it is useful to be able to illustrate a general method with a simple example, such as a program to compute the sum of an arbitrary sequence of functions:

Illustration 3.3 *Sum of squares.*

```
ss1[n_Integer, a_]:=
(* Compute ss1 = sum a[i], for i = 1, 2, ... n *)
Module[{sum = 0, i},
       For[i = 1, i <= n, i++, sum += a[i] ];
       sum];

f1[i_]:=i*i; ss1[3,f1]  14
```

Now we can compute for instance the sum of the harmonic series:
f2[i_]:=1/i; ss1[3,f2]   11/6.

Returning to the example of sums of squares, if we use recurrence relations for each term of this sum, we can write this program without any multiplications using the fact that $(i+1)^2 = i^2 + (2i+1)$.

Illustration 3.4 *Sum of squares without multiplications.*

```
ss1[n_Integer]:=
  (* Compute ss1 = sum i*i, for i = 1, 2, ..., n
     without multiplications *)
  Module[{sum = 0, i, odd = -1, term = 0},
         For[i = 1, i <= n, i++,
             odd += 2; term += odd; sum += term];
         sum];
```

```
ss1[3]   14
```

If the time to execute multiplication is longer than the time to execute two additions and two replacements, the second program is faster than the first. Unfortunately it is not obvious that the new program gives the correct results. In order to see how it works, let's make a table of the consecutive values of the variables inside the loop.

i	odd	term	sum
-	−1	0	0
1	1	1	1
2	3	4	5
3	5	9	14

Such a tabular arrangement used to systematically check if a program gives the correct results (at least for the data at hand) is called a **trace** of the program. We use it often to understand how a program works. Obviously good results of a program's trace do not prove its correctness, but it helps to find some obvious errors, often called **bugs**. Constructing a trace is an important part of "debugging" programs.

The best way to ensure a program's correctness is a proof. In most cases proving correctness is difficult. In our case it might look as follows:

We start our last program with the definition of four local variables: `sum=0`, `i`(counter), `odd=−1` and `term=0`. Next we have the **For** loop, and finally we deliver the result. I assert that for consecutive `i=0,1,...,n` values of the variables `odd`, `term`, and `sum` on entry to the loop satisfy the conditions `odd`=$2i-1$, `term`=$i*i$, and `sum`=$0*0+1*1+...+i*i$.

The proof is by induction. The assertion, as you can check, is true on entry, for `i=0`. Assuming the truth for an arbitrary integer `i`, we check them for `i+1`. The loop ends with `i=n` with the correct result, provided that the result is computable, i.e., does not cause overflow.

Such a set of relations that are true on entry to the loop, on exit, and deliver the desired result, is called the **invariant** of the loop.

Repetitions of a sequence of orders is a fundamental operation. Without such repetitions computers would be useless. In functional programs this repetition is achieved by multiple calls of the same function (with different arguments) until it gives the final result. Let us consider a program to compute factorial written in the functional style:

Illustration 3.5 *Functional factorial.*

```
factorial[n_]:=
  (* Compute n! in functional style. *)
  If[n < 2, 1, n * factorial[n-1]];

factorial[4]    24
```

Let us study the computation of 4! in this example. We start with a call to `factorial[4]`. The call goes to the second alternative of the **If**, and computes `4*factorial[3]`. The sequence of computations looks as follows:

$$factorial[4] = 4 * factorial[3] = 4 * 3 * factorial[2] = 4 * 3 * 2 * 1 = 24$$

In general a repetition in the functional style has two parts:

1. **Calls**: repeated calls of the same function until we know the result (in our case `factorial[1]`), and

2. **Returns**: when we finish each call and obtain a result.

Obviously the number of calls is the same as the number of returns. Such call of a function inside itself we call a *simple recursion* (a case of general recursion in which we call many functions, which may call the first one). This simple recursion is used for repetitions in functional programs.

Let us write the same program in an imperative style:

Illustration 3.6 *Imperative factorial.*

```
factorialImp[n_Integer]:=
  (* Compute n! in imperative style. *)
  Module[{product = 1, count},
        For[count = 2, count <= n, count++,
            product *= count ];
        product];

factorialImp[4]  24
```

In this program we also start with `factorialImp[4]`, but now everything is different. We define two local variables: *product*, for gathering successive products, and *count*, which will count how many times to repeat the loop. The computation proceeds by **replacements** of values in the pair of variables (*count*, *product*). The sequence of computations at the end of the **For** loop is

$$(2,2),\ (3,6),\ (4,24).$$

At the end of this sequence the **For** loop is finished, and the result is the value of the variable *product*, at this moment 24. The function is called only once, and there is no recursion.

If we call the program with the value of $n < 2$, the loop is omitted and the result is the value of *product*, which was initialized with the number 1.

Chapter 4

In the footsteps of George Boole,
who in 1854 described the elements of logic in his book "The Laws of Thought"

(Elements of propositional calculus)

Previously we introduced two pseudo-functions (spells) **If** and **For**, which we use to change the sequential order of computation depending on the value of a boolean <condition> (True or False). These boolean expressions have the form of propositions and can be quite complicated. Therefore, we now introduce the elements of propositional calculus more formally. A proposition is a sentence S that can be either true or false (also written 'True' or 'False' in our programs). For example 'Today is Monday' is a proposition, since one day a week it is true and the rest of the week it is false. Sentences such as 'Give me this book.' or 'How are you?' are not propositions. A proposition that is always true is called a *tautology*; a proposition that is always false is called a *contradiction*.

We will deal only with the operations on propositions that we use in our programs. We will use C-like notation. Some basic operations on propositions:

1. **Negation.** The simplest operation on propositions is *negation*, which we denote by '!'. Let $\mathcal{P}$ be a proposition then !$\mathcal{P}$ ("not $\mathcal{P}$") is called the *negation* of $\mathcal{P}$ and changes the value of $\mathcal{P}$ from True to False or False to True.

2. **Equivalence.** We say that two propositions $\mathcal{P}$ and $\mathcal{Q}$ are *equivalent* if they have the same value: either both are True or both are False given the same facts. For example 'This water is boiling' and 'The temperature of this water is above $100^\circ C$' are equivalent. The propositions 'n is a natural number' and 'n is an integer' are not, because mathematicians define natural numbers as the positive integers $1, 2, 3, \ldots$. We denote equivalence by '==' and write $\mathcal{P} == \mathcal{Q}$.

3. **Disjunction.** Let $\mathcal{P}$ and $\mathcal{Q}$ be propositions. Then $\mathcal{P} \parallel \mathcal{Q}$ (read '$\mathcal{P}$ or $\mathcal{Q}$') is also a proposition, called the *disjunction* or *logical sum* of $\mathcal{P}$ and $\mathcal{Q}$. A disjunction is True if any of its terms is True. For example $(10 > 20) \parallel (10 < 20)$ is True, since $10 < 20$ is True.

   We notice that the colloquial word 'or' may have a different meaning from our logical or. To illustrate, let us repeat an old joke. A computer scientist became a mother. Her good friend, upon learning the news, asked her: 'Girl or boy?' Our scientist answered: "Of course, what else could it be!?"

   To ruin this joke, let us set $\mathcal{P}$ to 'The baby is a boy'. Then !$\mathcal{P}$ is 'The baby is not a boy', which in our world means that the baby is a girl. The proposition $\mathcal{P} \parallel !\mathcal{P}$ can be false only if both $\mathcal{P}$ and !$\mathcal{P}$ are false, which is impossible, therefore $\mathcal{P} \parallel !\mathcal{P}$ is a tautology.

4. **Conjunction.** Again, let $\mathcal{P}$ and $\mathcal{Q}$ be propositions. Then $\mathcal{P}$ && $\mathcal{Q}$ (read '$\mathcal{P}$ and $\mathcal{Q}$') is a proposition called the *conjunction* or *logical product* of $\mathcal{P}$ and $\mathcal{Q}$. A conjunction is True

only if both $\mathcal{P}$ and $\mathcal{Q}$ are True. For example, the proposition $(0 < x)$ && $(x < 6)$ is True for x in the interval $0 < x < 6$, and False outside of this interval.

In propositional calculus as in arithmetic the order of evaluation of operations in an expression is important. To be consistent with most programming languages, we define the order of evaluations as follows: negation has the highest priority; then logical product (conjunction); logical sum (disjunction) has the lowest priority. In expressions where both logical and arithmetic operations are present arithmetic ($+, -, *$ and $/$) is done first, then relations ($<, >, \leq, \geq, ==$) and finally logical operations. These priorities are used in the C-like languages. Other languages, such as Pascal, use different priorities.

It is convenient to present results of logical operations in tabular form, called truth tables. In truth tables, we use T for True and F for False. We list all possible combinations of the propositions' values, then list the values for the expressions we are interested in. Below we give the truth tables for the four operations discussed above.

$\mathcal{P}$	!$\mathcal{P}$
T	F
F	T

Negation

$\mathcal{P}$	$\mathcal{Q}$	$\mathcal{P} == \mathcal{Q}$
T	T	T
T	F	F
F	T	F
F	F	T

Equivalence

$\mathcal{P}$	$\mathcal{Q}$	$\mathcal{P} \| \mathcal{Q}$
T	T	T
T	F	T
F	T	T
F	F	F

Disjunction

$\mathcal{P}$	$\mathcal{Q}$	$\mathcal{P}$&&$\mathcal{Q}$
T	T	T
T	F	F
F	T	F
F	F	F

Conjunction

As we can see from this table, equivalence, conjunction and disjunction are symmetric and transitive. We call a binary operation 'op' symmetric if
$$\mathcal{P} \text{ op } \mathcal{Q} == \mathcal{Q} \text{ op } \mathcal{P}.$$
We call it transitive if from
$$(\mathcal{P} \text{ op } \mathcal{Q}) \text{ \&\& } (\mathcal{Q} \text{ op } \mathcal{S}) \text{ it follows that } (\mathcal{P} \text{ op } \mathcal{S}) \quad .$$

We can use truth tables to prove formulae of the form $L == R$. For this we construct the truth table for the left hand side L and the table for the right hand side R and check if the entries in their result columns are identical for every combination of pairs (T,F). As an example, here are De Morgan's laws concerning the negation of conjunction and disjunction, namely:

$$!(\mathcal{P}\ \&\&\ \mathcal{Q}) == !\mathcal{P}\ ||\ !\mathcal{Q} \qquad !(\mathcal{P}\ ||\ \mathcal{Q}) == !\mathcal{P}\ \&\&\ !\mathcal{Q}\ . \qquad (4.1)$$

Below is a proof of the second law:

$\mathcal{P}$	$\mathcal{Q}$	$!(\mathcal{P}\ \|\|\ \mathcal{Q})$	$!\mathcal{P}$	$!\mathcal{Q}$	$!\mathcal{P}\ \&\&\ !\mathcal{Q}$
T	T	F	F	F	F
T	F	F	F	T	F
F	T	F	T	F	F
F	F	T	T	T	T

We used the truth tables of negation, conjunction, and disjunction, and noticed that columns three and five are identical.

We could end our discussion with some examples, but for a troublesome fact: the majority of computer languages, and in particular the ones from the C family, do not use the boolean algebra as we have described. Let us consider the following program fragment:

Illustration 4.1 *Example of If and boolean algebra.*

```
badIf[x_] := If[x > 0 && 1/x > 2, 1, 2];

badIf[1/3]   1
badIf[0]     2
```

We do not have any problem with the first answer: when we substitute $x = 1/3$, we obtain $1/3 > 0\ \&\&\ 3 > 2$ which is true, therefore 1 is the correct answer. But when we substitute $x = 0$, we have a problem since the number $1/0$ does not exist. Therefore, there is no way to compute this boolean expression. Nevertheless the machine has no trouble; it gives us the answer False, and prints the number 2.

For the languages in the C family, if the first atomic proposition is False, the whole conjunction is False and the second atomic part is not computed at all! Such operations are called conditional and the machine used the conditional 'and.' Disjunction has a similar property: if the first part of a disjunction is True, the second part is not computed, and the whole disjunction is True. Therefore, the machine used the conditional 'or.'

Both operations are neither symmetric nor transitive. This fact has serious consequences in programming. For efficiency, the first part of conditional operations should be short in execution, since it is always going to be executed, while the second part might be skipped.

Boolean operations are often used to check the computability of a given function. We call such checking **validation** of parameters. As an example, we will rewrite the previous program for the sum of squares:

Illustration 4.2 *Sum of squares.*

```
ss[n_Integer] :=
(* Compute sum ss = suma i*i, for i = 1, 2, ... n *)
Module[{nmax = 1860, sum = 0, i},
       If[n > nmax, 0,
          For[i = 1, i <= n, i++,
              sum += i * i];
       sum]];

ss[3]  14
```

How large is the n for which this sum exists? In our case

$$ss[nmax] < maxint = (2^{31} - 1).$$

If we do not know how to estimate the $nmax$, we must write a search program using a floating point sum (with less accuracy, but no integer overflow). In many cases the Euler-Maclaurin formula that associates a sum with an integral helps:

$$\sum_{k=1}^{n} f_k \approx \int_0^{n+1} f(k)\ dk - 0.5(f_0 + f_{n+1}) \qquad (4.2)$$

In our case, we have

$$\sum_{k=1}^{n} k^2 \approx \frac{(n+1)^3}{3} - \frac{(n+1)^2}{2} \qquad (4.3)$$

Using the first term we obtain the equation for the approximate $nmax$

$$\frac{(n+1)^3}{3} = 2^{31} - 1$$

which gives $n \approx 1859.73$.

The Euler-Maclaurin formula has an interesting geometric interpretation.

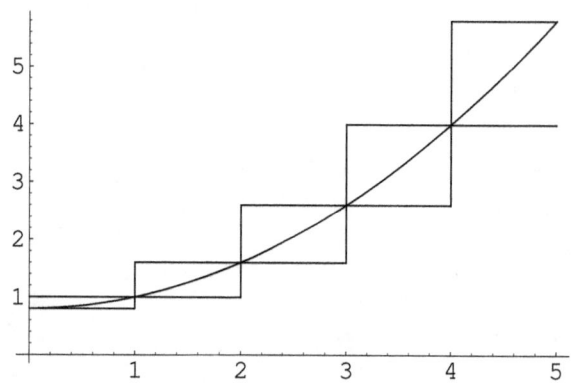

If we treat the integral

$$s = \int_0^{n+1} f(x)dx$$

for the increasing function $f(x)$ as the area under the function from $f(x)$ to the x axis, then

$$s_n = f(0) + f(1) + \cdots + f(n)$$

represents the area of the rectangles under the curve and

$$S_n = f(1) + f(2) + \cdots + f(n+1)$$

area of rectangles above this curve. Therefore

$$s_n \leq s \leq S_n \quad .$$

Taking the average $(s_n + S_n)/2$, gives the Euler-Maclaurin formula.

One interesting application of boolean algebra is to combine a binary representation of numbers and a mathematical representation of the boolean constants 1 for True and 0 for False. This way boolean constants are integers. C (and its generalization C++) is one of the languages that uses this convention extensively. Such languages have many operations that treat any integer as a sequence of bits and therefore execute very quickly. Below we list some such bitwise operations, and give their C notation in parentheses:

- BitNot[<int>] (~<int>)

- BitOr[<int1>,<int2>] (<int1> | <int2>)

- BitAnd[<int1>,<int2>] (<int1> & <int2>)

- LShift[<int>,<number>] (<int> << <number>) gives fast multiplications by powers of 2

- RShift[<int>,<number>] (<int> >> <number>) gives fast integer division by powers of 2

The instructions, which use the boolean expressions **If** and **For** in the C languages, also use the zero-one convention. These languages treat data as a sequence of bits, and are very liberal in conversion from one type into another. The boolean constants ('True' and 'False') and character constants are treated as integers, therefore we can use them in operations with type double. The expression

$$(1 > 0) + 'a' + 2.1$$

is legal in spite of the fact that 1>0 gives True, 'a' is a character, and 2.1 is a double.

We say that C-like languages use **weak types**, because they treat information as a sequence of bits, and do not demand explicit change of types to perform operations. This treatment might be useful in system programs, but can also be error prone. We note that *Mathematica* and *Java* do not allow mixing boolean and numeric types.

As an example consider a problem posed by Collatz in 1937: given a natural number n, construct a sequence C_i of natural numbers starting with $C_0 = n$ and for $i = 0, 1, 2, ...$

$$C_{i+1} = \begin{cases} 1, & \text{for } C_i = 1; \\ C_i/2 & \text{for } even\ C_i; \\ 3C_i + 1 & \text{for } odd\ C_i; \end{cases} \quad (4.4)$$

For example starting with $n = 3$ we obtain the sequence:

$$3,\ 10,\ 5,\ 16,\ 8,\ 4,\ 2,\ 1$$

It is conjectured that starting with any natural number this sequence terminates with 1. There is no proof for this conjecture, and there are numerous monetary awards for a proof. In the meantime, we can find many of the sequence's properties by programming.

We call the length of this sequence for a given n the *cycle-length*, where length means the number of elements in the sequence from n to its ending 1. In our example above, the cycle-length is 8.

One problem we can pose is finding the n of the longest sequence in a given interval $a \leq n \leq b < 10^6$. Let us write a program to find the length $len[n]$ of the Collatz sequence C_n:

Illustration 4.3 *Length of the Collatz sequence.*

```
col[n_Integer] := If[OddQ[n], 3n + 1, n/2];

len[n_Integer ? Positive] :=
   (* Compute the length of the Collatz sequence *)
   Module[{i, cc = n},
       For[i = 1, cc != 1, i++,
           cc = col[cc]];
       i ];
```

```
len[3] 8
```

Using bitwise operations defined above, notice that `OddQ[n]` can be replaced with `BitAnd[n,1]`, $n/2$ with `RShift[n,1]`, and even $3n + 1$ with `LShift[n,1]+n+1`.

Also notice that the function *col* is used only once, so we can substitute the code for it right into the body of the *len* function.

A careful reader will notice that we check the lower bound of n using `Integer ? Positive`. What is its upper bound? We have to find an n such that when we compute the sequence C_n no number will be greater than

$$maxint = 2^{31} - 1 = 2,147,483,647.$$

We need to know the maximum C_n, and for what n it happened. In our example $n = 3$, $maximum = 16$. Writing a short program, we can establish that such a maximum is achieved for $n = 113383$ and is $2,482,111,348 > maxint$.

I leave the details of the program as an exercise, and concentrate on some speedup of the program itself. You may ask why such speedup could be useful. In many applications, such as monitoring medical devices, computer games, or in programs controlling musical performance, execution time is critical. The computer in my pacemaker waits a few milliseconds for a beat of my heart and then sends an electrical signal. Any delay can cause fainting and even death. Therefore we have to learn how to speed up programs.

One source of slowness is counting by 1. Can we speed it up? Notice that if n is odd, $3n + 1$ must be even, so in this case we can go by 2 steps. The program might look as follows:

Illustration 4.4 *Length of the Collatz sequence revised*

```
len1[n_Integer ? Positive ] :=
(* Length of Collatz sequence with skip *)
   Module[{i, cc = n},
        For[i = 1, cc != 1, i ++,
             cc = If[OddQ[cc],
                    i++; cc + (cc+1)/2,
                    cc/2];
        ]
        i ];
```

len1[3]   8

Notice, that by changing $(3cc+1)/2$ to $cc + (cc+1)/2$, we increased the applicability of the program for integers $n > 113383$. I leave it as an exercise to compute the new $nmax$, and concentrate on further improving the program's time performance. We shortened the time of the loop by examining the last bit of n. Can we do better by examining 2 bits? Yes. For last 2 bits, we have the following 4 cases, using $k = Mod[n, 4]$ or $BitAnd[n, 3]$.

1. $n = 4k$; then skip=2, and the next C_n is $n/4$
2. $n = 4k + 1$; then skip=3, and the next C_n is $n - \frac{n-1}{4}$
3. $n = 4k + 2$; then skip=3, and the next C_n is $n - \frac{n-2}{4}$
4. $n = 4k + 3$; then skip=4, and the next C_n is $2n + \frac{n+5}{4}$

This interesting information helps our intuition to see how we can finish the C_n sequence with 1, since in all 3 cases our next (skipped) number is smaller, and only in the last case it is bigger.

We leave the implementation of this scheme as an exercise.

Chapter 5

In the footsteps of Donald Knuth,
who taught us mathematical analysis of programs

(Time complexity of programs)

What do we gain by writing programs in the imperative style? Up to now we did not show any advantages of imperative programs over functional programs. Even the program we presented in Chapter 3 computing a sum is simpler in the functional style, namely:

Illustration 5.1 *Functional sum of squares.*

```
ss2[n_Integer]:=
(* Compute sum of i*i in functional style. *)
  If[n < 1, 0, n*n + ss2[n-1]]

ss2[3]  14
```

One significant advantage of imperative programming is a difference in the speed of execution. In order to understand this, let us return to the computation of Fibonacci numbers.

$$f_1 = f_2 = 1, \quad f_k = f_{k-1} + f_{k-2}, \ k = 3, 4, \ldots \quad (5.1)$$

Let us consider the cost $K(n)$ of computation of f_n. The only long operation in this computation is the call to f_n. Therefore the cost, measured in time, is proportional to the number of such calls. Let us draw a tree of such calls for $n = 5$.

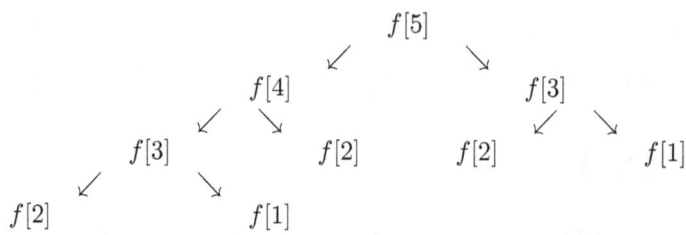

The cost K_n is a sum of the cost K_{n-1}, of the left branch of the tree, plus K_{n-2}, its right branch, plus 1, computation of the function itself. Together K_n gives the number of elements in the tree itself, for example $K_5 = 9$.
We have

$$K_1 = K_2 = 1, \quad K_k = K_{k-1} + K_{k-2} + 1. \quad (5.2)$$

Here is a table of f_n and K_n for the first few n.

n	1	2	3	4	5	6	7	8	9
f_n	1	1	2	3	5	8	13	21	34
K_n	1	1	3	5	9	15	25	41	67

A careful reader will notice that for each n in this table

$$K_n = 2f_n - 1 \quad (5.3)$$

To prove this is for every n, we substitute K_n from (5.3) into the equation (5.2). After simplification we obtain the same recurrence equation as for f_k, with the boundary conditions, therefore $K_n = 2f_n - 1$.

It is obvious that f_n and therefore K_n are increasing sequences. To see how quickly they increase, we solve a second order linear difference equation to obtain the explicit formula for f_n.

n	1	2	3	4	5	6	7	8	9
f_n	1	1	2	3	5	8	13	21	34
F_n	0.724	1.171	1.894	3.065	4.960	8.025	12.985	21.010	33.994

Let $f_n = C\lambda^n$ and substitute it into equation (5.1). We obtain the following equation for λ

$$\lambda^2 - \lambda - 1 = 0 \tag{5.4}$$

This quadratic equation has two solutions:

$$\lambda_1 = \phi = \frac{1+\sqrt{5}}{2} \approx 1.61803 \quad \text{and} \quad \lambda_2 = \frac{1-\sqrt{5}}{2} \approx -0.61803 \tag{5.5}$$

Therefore, as you can check, the general solution can be written as

$$f_n = C_1 \lambda_1{}^n + C_2 \lambda_2{}^n \tag{5.6}$$

where the constants C_1 and C_2 do not depend on n. We choose these constants to satisfy the boundary conditions $f_1 = f_2 = 1$ and obtain

$$f_n = \frac{1}{\sqrt{5}} \left[\left(\frac{1+\sqrt{5}}{2}\right)^n - \left(\frac{1-\sqrt{5}}{2}\right)^n \right] \tag{5.7}$$

The quantity $\lambda_2^n \approx (-0.61803)^n$ goes quickly to zero for large n, therefore

$$f_n \approx F_n = \phi/\sqrt{5} \approx 0.447214 * (1.61803)^n \tag{5.8}$$

as we can see in the previous table.

Since f_n is an integer, it follows from equation (5.8) that f_n grows exponentially with exponent 1.62... and the cost function K_n also grows exponentially with the same exponent.

For comparisons of the growth of the two functions $f(n)$ and $g(n)$ for large n, it is useful to introduce $\mathcal{O}(\ )$ notation (read 'big-oh notation'):

$$f(n) = \mathcal{O}\big(g(n)\big) \tag{5.9}$$

which we read as "$f(n)$ is of the order at most $g(n)$ if there exist constants C and N such that $f(n) \leq C \cdot g(n)$ for every $n > N$."

This notation is used to describe the **time complexity** of programs. Using it we can say that the program `Functional Fibonacci` computes these numbers in exponential time $\mathcal{O}((1.61803)^n)$. We will see later that Fibonacci numbers can be computed in linear time $\mathcal{O}(n)$, or even logarithmic time $\mathcal{O}(\log n)$.

Let us now write a program in imperative style to compute the same function. We start by computing a few Fibonacci numbers: $f_3 = f_2 + f_1 = 2$, $f_4 = f_3 + f_2 = 2 + 1 = 3$.

To compute the next number f_{n+1} we have to know two previous numbers f_n and f_{n-1}. Let us call these numbers (which we must remember) as fn and $fnm1$. We obtain the program:

Illustration 5.2 *Imperative Fibonacci numbers.*

```
fibImp[n_Integer]:=
(* Compute  Fibonacci number fibImp(n) *)
Module[{fn = 1, fnm1 = 1, i, temp, maxFib = 46},
       (* return 0 for undefined fibImp(n) *}
       If[n < 1 || n > maxFib,
          fn = 0,
          For[i = 3, i <= n, i++,
              temp = fn; fn += fnm1; fnm1 = temp];
          fn]
    ];

fibImp[5] 5
```

Here again we used the fact that the **For** loop is not executed if the $<condition>$ is not satisfied. We also validate the computability of f_n, and return zero when f_n is not computable. We compute $MaxFib$ using equation (5.8):

$$f_n \approx F_n \approx 0.447214 * (1.61803)^n = MaxInt$$

We get $MaxFib = 46$ for 32-bit numbers, and $MaxFib = 92$ for 64-bit natural numbers. So we cannot compute numbers greater than $MaxFib$. In functional programs we assumed that computer arithmetic is such that we can get arbitrarily large integers. In imperative programs we cannot make this assumption.

We also want to substitute in parallel

$$fn \leftarrow fn + fnm1, \ fnm1 \leftarrow fn$$

Since replacement is destructive, we use a variable *temp* to save fn before it gets replaced by the new value. We could instead have used a correction

$$fn = fn + fnm1; \ fnm1 = fn - fnm1$$

Then we would not need the *temp* variable. The use of this correction gives us a new idea about computing Fibonacci numbers. If, in the loop, we write

$$fnm1 = fnm1 + fn; \quad fn = fn + fnm1$$

then in one step we compute two consecutive Fibonacci numbers. Our modified program looks as follows:

Illustration 5.3 *Fibonacci numbers in a double loop.*

```
fibImp2[n_Integer] :=
(* Compute Fibonacci number f(n). Double loop. *)
  Module[{fn = 1, fnm1 = 1, i, maxFib = 46},
        (* return 0 for undefined fibImp2(n) *}
        If[n < 1 || n > maxFib, fn = 0,
           For[i = 2, i <= Quotient[n,2], i ++,
               fnm1 += fn; fn += fnm1];
           If[OddQ[n] && n != 1, fn += fnm1]];
        fn];

fibImp2[5]   5
```

This program is nearly two times faster than the previous one. Can we speed it up more? Before we do, let us consider how to compute the time complexity of imperative programs.

For functional programs, we counted the number of calls to the function. This number represented the number of repetitions. In imperative programs, repetitions are done using the **For** loop, so the number of repetitions of the loop expresses the program's time complexity. For our two last programs this number is $\mathcal{O}(n)$. As we see, such a measure is not sharp enough to catch subtle differences. We also notice that the only way that we can do replacement in functional programs is by substitution in parameters, so we can write our functional Fibonacci program in linear time also, namely:

Illustration 5.4 *Linear functional Fibonacci numbers.*

```
funAux[n_, fn_, fnm1_]:=
(* Auxiliary function for replacements *)
  If[n > 1, funAux[n-1, fn + fnm1, fn], fn]];

fibFun2[n_Integer]:=
(* Functional  Fibonacci number f(n) in linear time *)
  funAux[n, 1, 0];

fibFun2[5]   5
```

We start with a call to $funAux[5, 1, 0]$. Then there are the recursive calls $funAux[4, 1, 1]$, $funAux[3, 2, 1]$, $funAux[2, 3, 2]$, and $funAux[1, 5, 3]$. We go to the other branch of the **If** and return $fn = 5$. We have exactly n calls.

Now we return to systematically reducing repetitions in the computation of Fibonacci numbers. For this purpose we prove, by induction, the following theorem for the matrix F:

$$F = \begin{pmatrix} 1 & 1 \\ 1 & 0 \end{pmatrix}^n = \begin{pmatrix} f_{n+1} & f_n \\ f_n & f_{n-1} \end{pmatrix} \qquad (5.10)$$

For $n = 1$ the theorem is true, because $f_2 = f_1 = 1$ and $f_0 = 0$. Assume the theorem is valid for n. To prove its correctness for $n + 1$ we multiply both sides of (5.10) from the left hand side by the matrix

$$\begin{pmatrix} 1 & 1 \\ 1 & 0 \end{pmatrix}$$

We obtain

$$\begin{pmatrix} 1 & 1 \\ 1 & 0 \end{pmatrix}^{n+1} = \begin{pmatrix} f_{n+1} + f_n & f_n + f_{n-1} \\ f_{n+1} & f_n \end{pmatrix} = \begin{pmatrix} f_{n+2} & f_{n+1} \\ f_{n+1} & f_n \end{pmatrix}$$
$$(5.11)$$

which ends the proof.

Here we used the fact that to multiply two square $n \times n$ matrices $C = A \times B$ with elements a_{ij} and b_{ij}, we sum the products of rows by columns, i.e.,

$$c_{pq} = \sum_{k=1}^{n} a_{pk} b_{kq}$$

From equation (5.10) it follows that

$$f_{2n} = 2f_n f_{n+1} - f_n^2, \quad f_{2n+1} = f_{n+1}^2 + f_n^2, \quad f_{2n-1} = f_{n-1}^2 + f_n^2 \tag{5.12}$$

To prove this we use (5.10) with $n = k + m$ to obtain

$$\begin{pmatrix} 1 & 1 \\ 1 & 0 \end{pmatrix}^{k+m} = \begin{pmatrix} f_{k+m+1} & f_{k+m} \\ f_{k+m} & f_{k+m-1} \end{pmatrix}$$
$$= \begin{pmatrix} f_{k+1} & f_k \\ f_k & f_{k-1} \end{pmatrix} \times \begin{pmatrix} f_{m+1} & f_m \\ f_m & f_{m-1} \end{pmatrix} \tag{5.13}$$
$$= \begin{pmatrix} f_{m+1}f_{k+1} + f_m f_k & f_m f_{k+1} + f_k f_{m-1} \\ f_m f_{k+1} + f_k f_{m-1} & f_m f_k + f_{m-1} f_{k-1} \end{pmatrix}$$

and substitute $n = m = k$.

Now we are ready to show a 'fast' way to compute Fibonacci numbers f_n. The main idea is to organize the computations using two operations: 'collecting' and 'doubling'. We explain it using an example of power raising.

To compute x^N for an arbitrary x and natural N, we write the binary representation of N as $N = n_k 2^k + n_{k-1} 2^{k-1} + n_1 2^1 + n_0$, where $n_k = 1$ and every other n_i is 0 or 1. Then

$$x^N = x^{\sum_{i=0}^{k} n_i 2^i} = x^{2^k} \times x^{n_{k-1} 2^{k-1}} \times \cdots \times x^{n_1 2^1} \times x^{n_0} \tag{5.14}$$

We have two cases:

1. $n_i = 0$, then this term is equal to one and in multiplication can be skipped.

2. $n_i = 1$, then the product has to be multiplied by x^{2^i}.

We construct two sequences $D_i = x^{2^i}$ and $P_i = D_i \times P_{i-1}$ for $i = 0, 1, ..., k$. We write the following program (assuming that x^N exists):

Illustration 5.5 *Fast power computation.*

```
pow[x_, n_Integer?Positive]:=
(* Compute x^n fast *)
  Module[{k, col = 1, db = x},
        For[k = n, k > 1, k = Quotient[k,2],
            If[OddQ[k], col *= db];  (* collecting *)
            db *= db];   (* doubling *)
        db * col];
```

pow[2,5] 32

We use the fact that an odd natural number ends with 1 in the binary system. The suffix ?Positive of the parameter n_Integer ensures $n \geq 0$. The trace of a computation of x^{12} is

k	12	12	6	3
col	1	1	1	x^4
db	x	x^2	x^4	x^8

The second column gives the values at the start of the **For** loop and the other columns at its end. In a similar way we can multiply $N \times x$ using only addition.

The program to compute f_n numbers is very similar. In this case we have a pair of numbers $db = d1, d2$ for doubling, and a pair $pr = p1, p2$ for collection. The second number of the pair has a subscript of one less than the first one. Here is the program:

Illustration 5.6 *Fibonacci numbers in logarithmic time.*

```
fibFast[n_Integer?Positive]:=
(* Compute Fibonacci numbers $f_n$ fast *)
  Module[{k, p1 = 1, d1 = 1, p2 = 0, d2 = 0, temp},
        For[k = n, k > 1, k = Quotient[k,2],
            If[OddQ[k],
               temp = d1*p1 + d2*p2;
               p1=p2*d1 + p1*(d1+d2);
               p2=temp];    (* collection *)
            temp=d1*d1 + d2*d2;
```

```
        d1=d1*d1 + 2*d1*d2;
        d2=temp];     (* doubling *)
     d1*p1 + d2*p2];
```

`fibFast[6] 8`

The trace of this program for f_{12} is

m	12	12	6	3
d1	1	1	3	21
d2	0	1	2	13
p1	1	1	1	5
p2	0	0	0	3

The final result is $f_{12} = d1 \cdot p1 + d2 \cdot p2 = 21 \cdot 5 + 13 \cdot 3 = 144$.

Looking at this program, we are not convinced that it is fast! For simplicity, let us analyze the program for x^n.

The **For** loop of the program is executed exactly k times, where k is obtained from the equation $n/2^k = 1$, or $k = Floor(\log_2 n) + 1$, so the loop is done in logarithmic time. So there is such an N that, for $n > N$, our program will be faster than all linear programs. Later we will use similar programs to compute $x^n \bmod m$, Perrin numbers, and similar quantities.

Note we can skip the long formulae in the program if we write a program to multiply 2 matrices and know how to represent them in memory.

Finally, there is a natural question: where did we get the matrix F? Recall

$$F = \begin{pmatrix} 1 & 1 \\ 1 & 0 \end{pmatrix}$$

To answer this, we have to review some concepts from linear algebra.

We call a number λ an *eigenvalue* of a matrix A if there exists a vector x such that $Ax = \lambda x$. Since

$$Ax - \lambda x = Ax - \lambda I x = (A - \lambda I) x = 0$$

then x is a solution of a square homogeneous system, therefore the determinant

$$d = \det(A - \lambda I) = 0 \qquad (5.15)$$

has to be zero. We call (5.15) a characteristic polynomial of the matrix A.

Consider an $n \times n$ matrix A and let

$$B_A(\lambda) = \det(\lambda I_n - A) = \lambda^n + p_1 \lambda^{n-1} + \cdots + p_{n-1}\lambda + p_n$$

be its characteristic polynomial. Then the matrix

$$\mathbf{F}[\mathbf{A}] = \begin{pmatrix} -p_1 & -p_2 & -p_3 & \cdots & -p_{n-1} & -p_n \\ 1 & 0 & 0 & \cdots & 0 & 0 \\ 0 & 1 & 0 & \cdots & 0 & 0 \\ \vdots & \vdots & \ddots & \ddots & \vdots & \vdots \\ 0 & 0 & \cdots & 1 & 0 & 0 \\ 0 & 0 & 0 & \cdots & 1 & 0 \end{pmatrix}$$

has the same eigenvalues as the matrix A. It is also called the *Frobenius form* of A. For the proof we have to expand the determinant of $F[A]$ by the first row. Now you can see that the matrix

$$F = \begin{pmatrix} 1 & 1 \\ 1 & 0 \end{pmatrix}$$

is a Frobenius matrix, so its eigenvalues satisfy the same equation as the equation for Fibonacci numbers given by $f_n = c\lambda^n$ numbers, namely

$$\lambda^2 - \lambda - 1 = 0 \tag{5.16}$$

Chapter 6

In the footsteps of John Backus,
who in 1954 working for IBM invented FORTRAN, the oldest higher level programming language still used today

(Using data arrays)

In Chapter 3 we described von Neumann's machine with its numbered memory cells that can be addressed sequentially. Such a sequence of data of the same type having a common name we call a one-dimensional array. FORTRAN was the first language that used arrays to remember the data needed for future computations.

To define an array we give it a name and we may also define initial values of its elements. *Mathematica* has many constructors of arrays. One of them is `Table[<in>,{<num>}]`, where `<in>` gives the initial values of the array and `<num>` gives the number of its elements. For instance `fib = Table[1,{46}]` defines an array named `fib` of 46 elements, each element initialized with a number one, and `evennum = Table[2i,{i,5}]` constructs an array of five elements: 2, 4, 6, 8, 10.

Another way is to list its elements: `evennum = {2, 4, 6, 8, 10}` produces the same result. To choose an element in order to set or retrieve its value we use `<name>[[<which>]]`, where `<which>` is called an *index* into an array. For instance if `tab = {0,1}`, then `tab[[1]]=0` and `tab[[2]]=1`.

We illustrate the use of arrays in the computation of Fibonacci numbers.

Illustration 6.1 *Fibonacci numbers with array.*

```
fibArray[ n_Integer] :=
(* Compute a Fibonacci number using an array. *)
  Module[{maxFib=46, fib=Table[1,{46}], i, result},
         (* return 0 for undefined fibArray[n] *)
         If[n < 1 || n > maxFib, result = 0,
            For[i = 3, i <= n, i++,
                fib[[i]] = fib[[i-1]] + fib[[i-2]]
                ];
            result = fib[[n]] ];
         result
  ];

fibArray[12]   144
```

The array `fib` vanishes at the end of the operation of the `fibArray` function. The computed values of the array can only be used until the end of this function. They are addressed using the index of the array name, in our case `fib[[i]]`.

Variables that exist only inside a function are called *local variables* (sometimes *automatic variables*). They are independent of the same names outside the function. In some applications we would like to share the variables among different programs. We can use two methods:

1. Define a variable as *global*, which makes it visible to every function. We then lose control of the variable and the safety of its use, and face the possibility of an unexpected change by other programs.

2. Ask the compiler to leave a variable value or array for later use locally. We call such a non-vanishing object *static*.

We illustrate the sharing of a variable among functions in an example of lazy computation. Here *lazy* means to compute only those values necessary at the moment. This example also remembers the values for future use.

Illustration 6.2 *Fibonacci numbers with lazy computation.*

```
Begin["fibo`"]
  f = Table[1,{46}]; count = 2;
End[];

fibLazy[n_Integer]:=
(* Compute Fibonacci numbers the lazy way. *)
  Module[
    {maxFib=46,i},
    (* return 0 for undefined fibLazy[n] *)
    If[n < 1 || n > MaxFib, 0,
      If[n > fibo`count, (* Compute more numbers *)
        For[i = fibo`count+1; fibo`count = n, i <= n, i++,
          fibo`f[[i]] = fibo`f[[i-1]] + fibo`f[[i-2]]
        ]
      ],
      fibo'f[[n]]
    ]
  ];

fibLazy[12]  144
```

In the Footsteps of Programming Teachers

The function `Begin["<name>`"]` defines a new *context* called `<name>`. All names used in this context are invisible outside it, but they are static. To use them, one has to use their whole name, including the context name. The function `End[]` returns to the previous context. Notice how the number of valid elements grows only as needed, starting from 2 numbers (1,1) written at the beginning.

A similar technique is used in *dynamic programming*, in which we judge what information is needed, compute it only once, and store it to be picked up later.

In the next example, we compute the Fibonacci numbers only the first time, and then we use them.

Illustration 6.3 *Fibonacci numbers computed dynamically.*

```
Begin["fibo`"]
  f = Table[1,{46}]; first = True;
End[];

fibDynamic[n_Integer] :=
  (* Compute  Fibonacci number dynamically. *)
  Module[
    {maxFib=46,i},
    (* return 0 for undefined fibDynamic[n] *)
    If[n < 1 || n > MaxFib, 0,
       If[fibo`first,
          (* First time *)
          fibo`first = False;
          For[i = 3, i <= maxFib, i++,
              fibo`f[[i]] = fibo`f[[i-1]] + fibo`f[[i-2]]
          ];
          fibo'f[[n]]
       ]
    ];

fibDynamic[12]   144
```

The use of arrays simplifies the program to compute Fibonacci numbers in logarithmic time using matrix operations. Given two matrices

$$A = \begin{pmatrix} a11 & a12 \\ a21 & a22 \end{pmatrix}$$

and

$$B = \begin{pmatrix} b11 & b12 \\ b21 & b22 \end{pmatrix}$$

the scalar product is

$$P = A \cdot B = \begin{pmatrix} a11 \cdot b11 + a12 \cdot b21 & a11 \cdot b12 + a12 \cdot b22 \\ a21 \cdot b11 + a22 \cdot b21 & a21 \cdot b12 + a22 \cdot b22 \end{pmatrix}$$

If we assume all matrices used and their scalar products are symmetric, as is the case for matrices used in computing Fibonacci numbers, we can represent such matrices M as an array of three numbers: $\{m11, m12, m22\}$. The unit matrix will be $One = \{1, 0, 1\}$ and the matrix $F = \{1, 1, 0\}$. Here is the program to compute Fibonacci numbers using matrices:

Illustration 6.4 *Fibonacci numbers computed with matrices.*

```
mul[a_List, b_List]:=
  Module[{t},
        t=a[[2]] b[[2]];
        (* multiply two matrices *)
        {a[[1]] b[[1]] + t,
         a[[1]] b[[2]] + a[[2]] b[[3]],
         a[[3]] b[[3]] + t}]

fibMat[n_Integer?Positive]:=
(* Compute Fibonacci numbers fast *)
  Module[{k, col={1, 0, 1}, db= {1, 1, 0}},
        For[k = n-1, k > 1, k = Quotient[k,2],
            If[OddQ[k],
                col=mul[col,db]]; (* collecting *)
                db= mul[db, db]]; (* doubling *)
        mul[db, col][[1]]]

fibMat[12] 144
```

The possibility of accessing any element of an array makes it convenient to store information about arbitrary objects. This raises the interesting problem of searching for information: given an array of n elements $a_1, a_2, ..., a_n$ and an element x, find such an index i that $a_i == x$. Return 0 if x is not in the table. Our program might look as follows:

Illustration 6.5 *Sequential search in array.*

```
place[a_List, n_Integer?Positive, x_]:=
(* Search for x in array a *)
  Module[{i},
        For[i = 1, a[[i]] != x && i < n, i++, ];
            If[a[[i]] == x, i, 0]
        ];

place[{3,2,1},3,5]    0
place[{3,2,1},3,1]    3
```

Our element x may be of any type. After a strange loop with the $< action >$ missing, we do not know upon finishing whether we found our element or examined every element of the array without finding it. We must check again with the **If** statement. If the array is global, the **If** statement is not needed, and our program might be

Illustration 6.6 *Sequential search in global array.*

```
a = {0,1,2,3}; (* Initialize array *)
nNa = 4; (* Size of the array *)
place1[x_]:=
(* Search for x in global array a *)
  Module[{i},
        a[[1]] = x;
        For[i = nNa, a[[i]] != x, --i, ]; i-1];

place1[5]    0
place1[3]    3
```

In this program our array *a* is bigger by one than the information stored. The first place is used to store the search element *x*. We then search the array from the end. If the element is in the array, we find it before examining the first element. If not, we find it in the first place. So, the logic of the program is simpler.

The complexity of linear search is $\mathcal{O}(n)$. If the array is small, sequential search is fast, but we might put the most often searched for elements at the end of the array. If we do not know the frequency of usage, we could write a search program which also counts and keeps these counts for some time in "training" mode. Then we could use the information gathered to reorder our array.

For a large array, we should sort it and use binary search. Binary division of an ordered set is a strategy used in everyday life, even by children.

Imagine a game between a child and a computer in which the computer asks for a person's age or the price of a product. The child guesses a number and the computer can only give 3 answers: the number is greater, smaller, or "you got it right." Children quickly discover that if they can locate the number N they are looking for in an interval $(Nmin, Nmax)$ such that $Nmin < N < Nmax$, then the best strategy for finding N in the long run is to guess N in the middle.

Given a sorted array of size n, its middle is $1 + Quotient[n-1, 2]$. We replace either the left or right part of the $(1, n)$ interval by this number. Then we repeat this process for half of the original interval.

A simple version of binary search is the following program:

Illustration 6.7 *Binary search in an ordered array.*

```
placeB[a_List, n_Integer?Positive, x_]:=
(* Binary search in ordered array a. *)
  Module[{left = 1, right = n, mid},
        For[ , left <= right, ,
            mid = left + Quotient[right-left, 2];
            If[a[[mid]] <= x, left = mid + 1];
            If[a[[mid]] >= x, right = mid - 1]; ];
        If[a[[mid]]==x, mid, 0]]

placeB[{1,2,3,4,5,6,7},7,5]    5
```

In the Footsteps of Programming Teachers

We start with setting the `left` and `right` parts of the search. We test if there is still more space to search and `mid` is computed safely, to make sure that there is no overflow when the size of the array is large. When we have found our element `x` in the place `mid`, we have `left=mid+1` and `right=mid-1`, therefore `left>right`, and we end the loop. Notice that `Quotient[left+right,2]` may overflow, but `left+Quotient[right-left,2]` cannot.

The time complexity is $\mathcal{O}(\log_2(n))$ since the search interval is decreased by half each time the loop is repeated.

Binary search programs given in many programming books have errors because the computation of $left + right$ may overflow. I have seen a book in which the author rigorously proves the search program's correctness and, in spite of this proof, the program does not work. This sad fact shows that proof of program correctness is not an easy task.

Binary search does not have to search only arrays, or address equality. Here is an example: for an increasing function $f[n]$, find a natural number n such that for a given $x > 0$, we have $f[n] <= x < f[n+1]$. In our program we will use also an approximate guess g, and a new pseudo-function `While` for controlling a loop:

```
While[<cond>, <action>] == For[ , <cond>, , <action>]
```

Illustration 6.8 *Binary search for an inverse function.*

```
searchN[x_, f_, g_]:=
(* Find natural n such that f[n]<= x <f[n+1] *)
(* for an increasing function f[n] *)
(* g is an approximation *)
  Module[
    {le, ri, mid = 1, nx = N[x], try},
    try= f[g];
    If[try<nx,
       le = g;
       While[try<nx,
             le*=2;
             try=f[le]];
       le/=2;
       ri=Ceiling[2le]+1;,
       ri=g;
```

```
        While[try>=nx,
              ri=Ceiling[ri/2];
              try=f[ri]];
        le=Floor[ri];
        ri=Ceiling[2ri]+1;
    ];   (* Binary search *)
    While[ri-le>1,
          mid=Floor[le+(ri-le)/2];
          If[f[mid]<=nx, le=mid, ri=mid]
    ];
    le]
```

The program consists of three parts: the search for the right boundary, when the value of try is too small; the search for the left boundary, when try is too large; and finally, the binary search when we already found an appropriate interval. We also use two functions; Floor[x] and Ceiling[x] for smaller and larger values of the integer part of x. However this program is very dangerous because the search might easily go beyond the domain of the function $f[n]$ causing an overflow. Here are examples of its application:

Illustration 6.9 *Compute the integer part of $\sqrt{12345678}$.*

```
sq[x_]:= x*x;
searchN[12345678, sq, 3000]

3513
```

Illustration 6.10 *How many factorials can be represented in 32 bits?*

```
fac[x_]:= x!; searchN[2.^(31)-1, fac, 8]
12
```

Illustration 6.11 *How many sums of squares can be represented in 32 bits?*

```
sms[n_]:= Sum[i*i,{i,n}]; searchN[2.^(31)-1, sms, 990]
1860
```

Chapter 7

In the footsteps of Isaac Newton,
who taught us how to raise a sum to the n-th power

(Binomial numbers)

Binomial numbers are used in many applications. In particular, they are very useful in combinatorics, where the binomial coefficient $\binom{n}{m}$ gives the number of ways in which a set of n objects can be divided when we take m of them at a time. We can define and compute $\binom{n}{m}$ from the formula:

$$\binom{n}{m} = \frac{n!}{m!(n-m)!} \qquad (7.1)$$

where $n! = 1 \cdot 2 \cdots n$. For example $\binom{4}{2} = \frac{4!}{2!2!} = \frac{4 \cdot 3}{2} = 6$, since there are six ways we can choose two elements from a set of 4 elements: $(1,2), (1,3), (1,4), (2,3), (2,4), (3,4)$.

Here is a program to compute $\binom{n}{m}$:

Illustration 7.1 *Binomial numbers using factorial.*

```
binomial[n_Integer, m_Integer]:=
(* Compute binomial number using factorial. *)
  If[n < 0 || n < m || n > MaxFac,
    (* Return 0 for undetermined binomial *)
    0,
    factorial[n]/ factorial[n-m]/ factorial[m]]
```

We leave the program to compute the factorial as an exercise for the reader. Notice that factorial grows so fast that very few of them can be computed. Since

$$MaxInt = 2^{31} - 1 \approx 2.14 \cdot 10^9$$

for 32-bit integers and

$$MaxLong = 2^{63} - 1 \approx 9.22 \cdot 10^{18}$$

for 64-bit integers, MaxFac has to be 12 (for 32-bit numbers) and 20 (for 64-bit numbers).

Knowing that

$$\binom{n}{k+1} = \binom{n}{k} \cdot \frac{n-k}{k+1} \text{ and } \binom{n}{0} = 1 \quad (7.2)$$

we can compute more combinatorics $\binom{n}{k}$ using this formula. To demonstrate let us rewrite (7.2) as

$$\binom{n}{k} = \frac{n(n-1)(n-2)\cdots(n-k+1)}{1 \cdot 2 \cdot 3 \cdots k} \quad (7.3)$$

If we write the binomial program carefully: $n \cdot (n-1)/2 \cdot (n-2)/3$, we notice that $n(n-1)$ is always divisible by 2 and $n(n-1)(n-2)$ is divisible by 6, so we can use integer division (obtaining integers).

We can prove (7.2) and (7.3) using the definition (7.1).

We can reduce number of multiplications and divisions in the computation, noticing that

$$\binom{n}{k} = \binom{n}{n-k} \qquad (7.4)$$

so the upper limit of repetitions of (7.2) for $k = 0, 1, \ldots$ is $min[k, n-k]$.

The important question is: for which pairs n, k can we compute $\binom{n}{k}$?

From (7.2) we see that for a given n, $\binom{n}{k}$ is biggest for $k = n/2$, so let us study how big this number is.

$$\binom{2m}{m} = \frac{(2m)!}{(m!)^2} \qquad (7.5)$$

We estimate $n!$ using Stirling's formula which says that for large n

$$n! \approx st = \sqrt{2\pi} n^{n+1/2} e^{-n} \qquad (7.6)$$

It can be proved using the Euler-Maclaurin summation formula for

$$\log(n!) = \sum_{k=2}^{n} \log(k) \qquad (7.7)$$

To see how good this approximation is we compare a small table of factorials:

n	1	6	11	16	21
$n!$	1	720	$3.99 \cdot 10^7$	$2.09 \cdot 10^{13}$	$5.11 \cdot 10^{19}$
st	0.92	710.	$3.96 \cdot 10^7$	$2.08 \cdot 10^{13}$	$5.09 \cdot 10^{19}$

Using Stirling's formula we get

$$\binom{2m}{m} = \frac{(2m)!}{(m!)^2} \approx \frac{4^m}{\sqrt{\pi m}} \qquad (7.8)$$

Using (7.8) we see that $\binom{33}{16} \approx 1.2 \cdot 10^9$ can be represented as a 32-bit integer, but $\binom{34}{17} \approx 2.3 \cdot 10^9$ cannot. Similarly, $\binom{66}{33} \approx 7.2 \cdot 10^{18}$ can be represented as a 64-bit integer, but $\binom{67}{33} \approx 1.4 \cdot 10^{19}$ is already too large.

In the Footsteps of Programming Teachers

One of the most common way of computing binomial numbers is using Pascal's triangle. For this computation we use the formula

$$\binom{n}{k} = \binom{n-1}{k-1} + \binom{n-1}{k} \qquad (7.9)$$

that can be proven using the definition (7.1). The name "Pascal's triangle" comes from the way Pascal arranged the computation, namely

```
       1
      1 1
     1 2 1
    1 3 3 1
   1 4 6 4 1
```

We start each row with 1, and we finish with 1. Every other number in the row is obtained by adding the two adjacent numbers above it.

Using a simple array and assuming $maxbin = 33$, we obtain the following program:

```
binomPascal[n_Integer, m_Integer]:=
(* Compute binomial number using arrays. *)
  Module[{maxbin = 33, dT = Table[1,{34},{34}], i, j},
      If[n < 0 || n < m || n > maxbin,
          (* give 0 for undefined binomials *)
          0,
          For[i = 0, i <= n, i++,
              For[j = 1, j < i, j++,
                  dT[[i+1,j+1]] = dT[[i,j]] + dT[[i,j+1]]
              ]
          ];
          dT[[n+1,m+1]]
      ]];

binomPascal[4,2]  6
```

Notice that a[[i,j]] is an element of the two-dimensional array previously defined by Table[1,{34},{34}].

This program wastes storage and also computes nearly half of the elements unnecessarily, but it is very simple. It translates the function of two variables into an array. These types of programs are useful. They are easy to write, they provide a way to test more sophisticated programs, and they usually run fast. In this case the above program has an additional advantage: it is easy to change into an excellent dynamic program, which we leave as an exercise.

For those of you who want to use less memory, notice that in Pascal's triangle the next row depends only on the previous one, so in our computations we can keep only one row in memory. We put a line of ones at the beginning of the program and then overwrite the ones with correct results. For instance let us take the fourth row

$$1, 4, 6, 4, 1, 1, 1, ..., 1$$

To compute the fifth row we start from the fifth place in the array so as not to destroy needed elements, obtaining

$$1, 5, 10, 10, 5, 1, 1, ..., 1$$

Here is the program:

Illustration 7.2 *Binomials with a small array.*

```
binPascSmall[ n_Integer, m_Integer]:=
(* Compute the binomial coefficient using less memory. *)
  Module[{maxBin=33, i, j, dT = Table[1,{34}]},
      For[i = 1, i <= n, i++,
          For[j = i-1, j > 0, j--,
              dT[[j+1]] += dT[[j]]
          ]
      ];
      dT[[m+1]]
  ];

binPascSmall[5,2] 10
```

I leave as an exercise a program using formula (7.4) (symmetry) to reduce the number of steps in the loops, and the validation of the arguments **n** and **m**.

Chapter 8

In the footsteps of James Wilkinson, *who taught us analysis of numerical errors*

(Analysis of numerical errors)

Programming with real numbers presents additional troubles. Not only do we have the maximum limit maxReal for type float and double, but nearly every real number has limited accuracy. Imagine a car tachometer: as we drive, the mile counter adds up until we pass the upper limit, usually 999999. After that, the counter starts again from zero and shows an incorrect number of miles. This kind of arithmetic is called a modular arithmetic and is used by machines. Negative integers are usually represented by two's complement, since machines use binary representation. We show two's complemented integers below using 3-bit representation, where the upper row contains the decimal representations and the lower row their binary equivalents.

-4	-3	-2	-1	0	1	2	3
100	101	110	111	000	001	010	011

Integers whose binary representation starts with a 1 are negative; numbers starting with a 0 are positive. Since $(-a) + a = 0$, if we use the simple addition table of binary numbers, we obtain overspill and a result 000. For instance, adding $(-2) + 2 = 110_2 + 010_2 = 1000_2$.

To change the sign of a number we flip its binary representation (changing all ones to zeros and zeros to ones) and then add one. We do the same for 8, 16 or 32-bit numbers. There exists a negative number that has no positive equivalent and behaves like 0 ($-a = a$). In our table it is the number $-4 = 100_2$.

Real numbers $\mathcal{R}$ (float and double) are represented differently from integers. Each $\mathcal{R}$ is composed of 3 parts: a sign, a weighted exponent, and a mantissa. In decimal representation the number 2/3 may look like

+ 50	66	66	66	67

Here + is the sign, 50 the weighted exponent (an exponent with weight 50, which is the middle of two digits' range), and the rest of the digits are the rounded mantissa. We use a weight, so the exponent can be negative or positive, and has the property that we can use integer arithmetic to compare 2 real numbers as if they were integers. In our example real numbers are represented as $X = 0.M \cdot 10^{e-w}$, where e is an exponent and w is a weight. To have a unique and most accurate representation each number is normalized, so it does not have initial zeros. The exception is the number 0.0: it has the sign 0 and the rest of the digits are also 0.

The method used most often to represent real numbers $\mathcal{R}$ is the IEEE form: for 32-bit numbers a float F is 1+8+23: sign- 1 bit, weighted exponent- 8 bits, and mantissa- 24 bits, where the first bit (1) is assumed, since a normalized number has to start with a bit 1. For a 64-bit double D we have 1+11+52. The ranges of F and D are

$$10^{-45} \leq |F| \leq 10^{39} \quad \text{i} \quad 10^{-323} \leq |D| \leq 10^{308} \tag{8.1}$$

As we can see there are more numbers $|r| < 1$ than $|r| > 1$, so it can happen that the number x exists, but $1/x$ does not. The accuracy of a computation can be estimated, noticing that

$$2^{-24} \approx 6. \cdot 10^{-8} \quad \text{and} \quad 2^{-52} \approx 2. \cdot 10^{-16} \tag{8.2}$$

We can illustrate the difference between the representation of integers and $\mathcal{R}$ by thinking about odometers. In the case of natural numbers, at the beginning we have many leading zeros and, as the numbers grow, the number of leading zeros gets smaller until finally we get overspill and the number changes sign. When we use the $\mathcal{R}$ representation the small integers such as 1.0 have many trailing zeros, because they are normalized. As the numbers grow, the number of trailing zeros gets smaller, but only the leading digits are accurate, and the numbers become approximate. When the exponent becomes too large we get overflow, and computations stop. If we go in the direction of zero and the exponent gets too small, the number suddenly becomes 0.

We can write short 'spy' programs to compute $n!$ or $1/n!$ using the $\mathcal{R}$ representation. Let us write such a small program to check the accuracy of floating point numbers.

Illustration 8.1 *Check the accuracy of machine representation.*

```
numberOfPowers[]:=
(* How many negative powers of 2 can we add to 1.0 *)
  Module[{n, d = 1.0, temp = 2.0},
      For[n = 0, temp != 1.0, n++,
          d /= 2; temp = 1.0 + d];
      n];
```

numberOfPowers[]   46

In the world of mathematics the condition $1.0 + 2.^{-n} == 1.0$, which we are checking above, cannot be met. In the machine world it is met very quickly, giving us an n such that $2.^{-n}$ is too small to influence the comparison $1.0 + 2.^{-n} == 1.0$.

The difference between these two worlds is important because we are used to applying mathematical laws such as the distributive or associative law that in computer mathematics are not true.

$$(a + b) + c \neq a + (b + c)$$

$$(a * b) * c \neq a * (b * c)$$

$$(a + b) * c \neq a * c + b * c$$

I leave examples illustrating this as an exercise.

To study this interesting world of computing, we will use a model which illustrates at least some of the secrets of computer arithmetic.

Let x be a real number and x^* its machine representation. We call the difference

$$E(x) = x - x^* \qquad (8.3)$$

the error in x. $E(x)$ is not always useful. If we measure the length of a room to an accuracy of 1mm we will be very happy, but if we measure the thickness of a hair with the same accuracy of 1mm, we may say that our measurement is horrible. In the real world we favor the big guys (the room) at the cost of the little guy (the hair). A much better measure of the error is the relative error $R(x)$:

$$R(x) = \frac{x - x^*}{x} = \frac{E(x)}{x}, \quad \text{for } x \neq 0 \qquad (8.4)$$

Using our previous discussion about representation of numbers $\mathcal{R}$, we also assume that

$$E(n) = 0, \quad R(n) = 0 \qquad (8.5)$$

for small integers n, even if written as floats (for example 2.0).

Studying the properties of numbers $\mathcal{R}$ we have bad and good news. The bad news is that usually we do not know the number x, we know only x^* its approximation. The good news is that we do not need as much accuracy of the error itself. It is enough to know one or two accurate digits, so the errors $R(x) = 2 \cdot 10^{-5}$ and $R(x) = 2.4 \cdot 10^{-5}$ may be treated as equal. Thus, we can use a linear approximation of the Taylor formula for 2 variables to study the error in the four arithmetic operations:

$$f(x + h, y + k) \approx f(x, y) + h f_x(x, y) + k f_y(x, y) \qquad (8.6)$$

where $h = x - x^* = E(x)$, $k = y - y^* = E(y)$ and $f_z(x, y)$ is the partial derivative with respect to the variable z.

If you are worried about such an inaccurate treatment, I can say that it is only a model and that quite a large portion of classical physics also tolerates inaccuracy by assuming linearity of some physical laws.

If we write equation (8.6) in our notation, we obtain

$$E(f(x, y)) = f_x(x, y) E(x) + f_y(x, y) E(y) \qquad (8.7)$$

This equation is good enough to obtain rules for errors for all four operations, namely

$$E(x \pm y) = E(x) \pm E(y) \quad \text{and} \quad R(x \pm y) = \frac{x}{x \pm y} R(x) \pm \frac{y}{x \pm y} R(y) \tag{8.8}$$

for addition and subtraction and

$$R(x \cdot y) = R(x) + R(y) \quad \text{and} \quad R(x/y) = R(x) - R(y) \tag{8.9}$$

for multiplication and division.

For a function of one variable we obtain

$$R(f(x)) = \frac{xR(x) \cdot f_x(x)}{f(x)} = E(x) \left[\log f(x)\right]'$$
$$R(1/x) = -R(x) \tag{8.10}$$
$$R(-x) = R(x)$$

As an illustration of (8.10) we will compute the error of

$$f_1 = \left(\frac{\sqrt{5} - 2}{\sqrt{5} + 2}\right)^3$$
$$f_2 = \left(\sqrt{5} - 2\right)^6 \tag{8.11}$$
$$f_3 = \left(9 - 4\sqrt{5}\right)^3$$
$$f_4 = 2889 - 1292\sqrt{5}$$

As we can see we have four different formulae to compute the same number. Students, when asked, without exception give f_4 as the best (the most accurate) formula for computation. Let us use our model and in particular (8.10) to estimate the error caused by each formula. The only approximate quantity is $\sqrt{5} \approx 2.2360$. Let us substitute $\sqrt{5} \approx 2.25 = 9/4$ and perform the calculations. We obtain

$$f_1 = \left(\frac{1/4}{17/4}\right)^3 = \left(\frac{1}{17}\right)^3, \ f_2 = \left(\frac{1}{4}\right)^6, \ f_3 = 0, \ f_4 = -18 \tag{8.12}$$

Let us set $\sqrt{5} = x$ and $E(x) = \epsilon \approx .014$. Then

$$f_1 = \left(\frac{x-2}{x+2}\right)^3, \quad R(f_1) = 3\epsilon \left(\log(x-2) - \log(x+2)\right)' = 3\epsilon \frac{4}{x^2 - 4} = 12\epsilon \tag{8.13}$$

Similarly

$$f_2 = (x-2)^6, \quad R(f_2) = \frac{6\epsilon}{x-2} = 6\epsilon(x+2) = 24\epsilon \tag{8.14}$$

$$f_3 = (9 - 4x)^3, \quad R(f_3) = \frac{-12\epsilon}{9 - 4x} = -12\epsilon(9 + 4x) = -200\epsilon \tag{8.15}$$

and finally

$$f_4 = 2889 - 1292x$$
$$R(f_4) = \frac{-1292\epsilon}{2889 - 1292x} = -2\epsilon \cdot 1300 \cdot 2900 = -7.5 \cdot 10^6 \epsilon \tag{8.16}$$

Our model says that our formulae are becoming worse and worse, and the formula to compute f_4 is horrible.

It is worth noticing that even many talented students and mathematicians choose f_4 as the best, which shows that the majority opinion is not supported by solid knowledge. This lapse has two sources:

1. Students do not think that their teacher will work so hard to spoil a good formula.

2. There is a common belief that the size of the error is proportional to the number of operations, and in f_4 we have only 2 operations: multiplication and subtraction.

A moment of thought, however, tells us that this cannot be true, since modern computers perform billions of numerical operations but send satellites into correct orbits, so error size cannot be proportional to the number of operations carried out.

To understand the behavior of computer arithmetic better, let us study formulae (8.9) and (8.8). Let us denote

$$M = \max\left(R(x), R(y)\right), \quad m = \min\left(R(x), R(y)\right) \tag{8.17}$$

Then it follows from (8.9) that the maximum error in multiplication and division cannot be greater than $2M$.

If we assume that we add two numbers $\mathcal{R}$ with the same sign, then

$$m \leq R(x+y) \leq M \ .$$

If we subtract two numbers with the same sign, we may lose all accuracy in one operation. This happens when the numbers are nearly equal.

Going back to the study of f_4, we notice that $1292\sqrt{5} \approx 2888.99983$, so in subtraction from 2889 we lose 8 decimal figures. The formulae f_i are worse and worse, because to obtain a small result we subtract bigger and bigger numbers. The answer $f \approx 0.00017307$ can be computed by getting rid of the bad subtraction and obtaining the formulae

$$f_5 = \frac{1}{\left(\sqrt{5}+2\right)^6}, \quad f_6 = \frac{1}{\left(9+4\sqrt{5}\right)^3}, \quad f_7 = \frac{1}{2889 + 1292\sqrt{5}} \quad (8.18)$$

I leave the full analysis of these formulae as an exercise.

Chapter 9

In the footsteps of Brook Taylor, *who in 1715 taught us how to study and calculate a power series*

(Calculation of power series)

We have already used the Taylor series a few times for the approximation of a square root ($\sqrt{1+z} \approx 1 + z/2 + \cdots$) and for error analysis of the four basic arithmetic operations. In this chapter we look at the application of the Taylor series in greater detail.

If a function $f(x)$ of one real variable x has $n+1$ derivatives at the point $x = a$, then such a function can be expanded into the following Taylor series:

$$f(x) = f(a) + f'(a)(x-a) + \frac{f''(a)(x-a)^2}{2!} + \cdots + \frac{f^{(n)}(a)(x-a)^n}{n!} + R_n \quad (9.1)$$

where R_n (called a remainder) can be expressed by different formulae and denotes the error caused by truncation of the series after n terms. We sometimes call this error a *truncation* error.

If for $n \to \infty$, $R_n \to 0$, then the series (9.1) is called *convergent*; if $R_n \to \infty$, it is called *asymptotic*. For $a = 0$ the Taylor series is sometimes also called the Maclaurin series. We are interested in the computational aspects of the Taylor series: a slowly convergent series is useless, and sometimes an asymptotic series might be very useful.

From a computational point of view finding the coefficients

$$a_n = f^{(n)}(a)(x-a)^n/n!$$

is of fundamental importance. To find the coefficients we will use the technique of operators, usually linear.

As an illustration of this technique consider the Taylor expansion of e^x around $x = 0$.

Illustration 9.1 *Expansion of* $\exp(x)$ *by the Taylor series.*
Since $\exp'(x) = \exp(x)$, to find the coefficients we can use an operator

$$L(y) \equiv y' - y = 0, \quad y(0) = 1 \qquad (9.2)$$

We assume the existence and uniqueness of the Taylor expansion. Since all derivatives exist at the point $x = 0$, we may also assume

$$y = \sum_{n=0}^{\infty} a_n x^n \qquad (9.3)$$

and substitute (9.3) into (9.2). Comparing the coefficients of x^n we obtain a recurrence relation for the coefficients a_n:

$$a_0 = 1, \quad a_n = a_{n-1}/n \text{ for } n = 1, 2, ... \qquad (9.4)$$

Since mathematicians guarantee the existence and uniqueness of this expansion, in whatever way we obtain our coefficients, they are correct.

Here is the program:

```
eTaylor[x_] :=
(* Compute exp(x), accurate to 5 figures,
   using the Taylor series. *)
   Module[{eps = 0.00002, term = 1., sum = 1., i},
        For[i = 1, Abs[term] > eps, i++,
            term *= x/i; sum += term];
```

    sum];

eTaylor[1.] 2.71828

Here are a few comments on this program:

1. A function **Abs** is necessary since for $x < 0$, the **For** loop would not terminate properly.

2. For $x < 0$, $eps = 0.00005$ is big enough (using d'Alembert's test) to get 5 significant figures, but for $x > 0$ we have to improve the estimate.

3. A test $Abs[term] > eps$ is not good for big $|x|$. To have 5 significant figures, the test should be $Abs[term/sum] > eps$, but such a test can be troublesome. For instance if $x = -1$, the partial sum of the series for e^x

$$e^{-x} = 1 - x + x^2/2 - x^3/6 \cdots$$

gives $1 - 1$, therefore the partial sum would cause division by zero. We can change the test into $Abs[term] > eps * Abs[sum]$ and in this case avoid division by zero.

We have to be careful of the odd functions, say

$$\sin[x] = x - x^3/3! + x^5/5! - \cdots + \cdots$$

with an expansion around $x = 0$. In this case our test would cause an infinite loop. Therefore, in general we should weaken the test and use

$$Abs[term] \geq eps * Abs[sum]$$

4. The function $exp[x]$ grows quickly, so x cannot be too large. For 32-bit $\mathcal{R}$ numbers $xmax \approx 88.7$, and the program eTaylor needs 146 terms in the computation.

5. For small negative x, say $x = -10$, the Taylor series, in spite of its absolute convergence, is worthless. In this case

$$e^{-10} = 1 - 10 + 100/2 - 1000/6 + \cdots$$

A very small (positive) number is obtained by subtraction of $-2755.73 + 2755.73$ for $n = 9, 10$, so we will lose at least 8 digits for $x = -10$. In this case, for e^x we know that

$$e^{-x} = \frac{1}{e^x} \qquad (9.5)$$

therefore we may use the Taylor series only for $x \geq -1$.

6. For e^x, we can reduce the interval substantially: for $x = n + a$, $e^x = e^n \cdot e^a$, and $|a| \leq 0.5$ or what is often used in libraries of mathematical functions $e^x = 2^{m+z}$.

As another example, we will expand the function $y = \arcsin(x)/\sqrt{1-x^2}$ around $x = 0$. Differentiating y, we obtain an equation

$$(1-x^2)y' - xy = 1 \qquad (9.6)$$

from which we get $a_1 = 1$ and $a_{2n+1} = 2n \cdot a_{2n-1}/(2n+1)$, or

$$\frac{\arcsin(x)}{\sqrt{1-x^2}} = x + \frac{2}{3}x^3 + \frac{2 \cdot 4}{3 \cdot 5}x^5 + \cdots + \frac{(2n)!!}{(2n+1)!!}x^{2n+1} + \cdots \qquad (9.7)$$

In the programming of limits the use of the Taylor series is priceless. Let us look at computing $\infty - \infty$. In many cases a Taylor expansion allows us to change a bad subtraction into a division.

Notice that if $f \to \infty$ then $(1/f) \to 0$. Therefore we have:

$$f - g = \frac{1}{1/f} - \frac{1}{1/g} = \frac{1/g - 1/f}{1/f \cdot 1/g} \qquad (9.8)$$

We changed $\infty - \infty$ into $0/0$. If we now use the Taylor series we often obtain a good computational formula. For example:

Illustration 9.2 *Compute $y = (1/x) - (1/\sin(x))$ around $x = 0$.*
Using formula (9.8) we obtain

$$\begin{aligned} y &= \frac{1}{x} - \frac{1}{\sin(x)} = \frac{\sin(x) - x}{x\sin(x)} = -\frac{x^3(1/3! - x^2/5! + x^4/7! \cdots)}{x^2(1 - x^2/3! \cdots)} \\ &= -x\frac{h(x)}{1 - x^2 h(x)} \end{aligned} \qquad (9.9)$$

This formula is very good for computations around $x = 0$. Notice that the division of the numerator and denominator by x^2 is necessary for using (9.9) in a program.

We also used the formula

$$\sin(x) = x - \frac{x^3}{3!} + \cdots + \frac{(-1)^n x^{2n+1}}{(2n+1)!} + \cdots \qquad (9.10)$$

which we can obtain from books or using the operator

$$L(y) \equiv y'' - y = 0, \quad y(0) = 0, \quad y'(0) = 1 \qquad (9.11)$$

Let us write a program to compute

$$h(x) = \frac{1}{3!} - \frac{x^2}{5!} + \cdots + \frac{(-1)^n x^{2n}}{(2n+3)!} + \cdots \qquad (9.12)$$

```
hTaylor[x_] :=
(* Compute h(x), accurate to 5 digits,
   using Taylor series for |x|< 1 *)
  Module[{eps = 0.00005, term = sum = 1./6, u =- x*x, i},
    For[i = 4, Abs[term] > eps, i += 2,
      term *= u/(i*(i+1)); sum += term];
    sum]
```

with proper validation of $|x| \leq 1$. We can also estimate the truncation error and write a program for

$$h(x) = \frac{1}{6}\left(1 + \frac{u}{20}\left(1 + \frac{u}{42}\left(1 + \frac{u}{72}\right)\right)\right) \qquad (9.13)$$

where $u = -x^2$.

Notice the grouping of terms in the equation (9.13) for the computation of a polynomial with coefficients a_i. Such a grouping is known as Horner's rule (William Horner 1830). In general, a polynomial

$$\begin{aligned} p_n(x) &= a_0 + a_1 x + a_2 x^2 + \cdots + u_n x^n \\ &= a_0 + x\left(a_1 + x\left(a_2 + \cdots + x\left(a_{n-1} + x\, a_n\right)\cdots\right)\right) \end{aligned} \qquad (9.14)$$

is computed using a recurrence relation

$$y_0 = a_n, \quad y_i = y_{i-1} x + a_{n-i} \quad \text{for } i = 1, 2, \ldots n. \qquad (9.15)$$

We can write it as a program fragment:

```
y = a[[n+1]];
For[i = n, i > 0, --i, y *= x + a[[i]]]
```

where we assume an array of coefficients a_i for $i = 1, 2, ..., n+1$ as given.

Chapter 10

In the footsteps of Archimedes,
who in ancient times computed the number π accurately

(The computation of π)

> "How I wish I could calculate pi..."
>
> C. Heckman on memorizing π. The number of letters in each word gives its next digit.

The computation of π has interested many famous mathematicians. By about 250 A.D. the Greek mathematician Archimedes had already computed π to four decimal figures using only the 4 arithmetic operations $(+, -, *, /)$ and square root extraction. His method was to compute the circumference of inscribed and circumscribed polygons in a circle of radius $r = 1/2$. Since he knew that the circumference of the circle was $2\pi r$, he could estimate π this way from above and from below.

He started with a hexagon, and carried out the calculations to 12, 24, 48 and 96 sides, obtaining $3\frac{10}{70} < \pi < 3\frac{10}{71}$. As a young student working with a computer in 1958, I decided to repeat his calculations and improve them.

I will apply his method to areas of polygons starting with squares instead of hexagons, leaving the repetition of Archimedes' calculations as an exercise.

We start the calculations of areas of inscribed and circumscribed squares on a circle of radius $R = 1$ and repeat them, doubling the number of sides each time.

Using trigonometry we can show that the value of π obtained this way satisfies the relation:

$$(N/2)\sin(2\pi/N) < \pi < N\tan(\pi/N) \qquad (10.1)$$

where N is the number of sides.

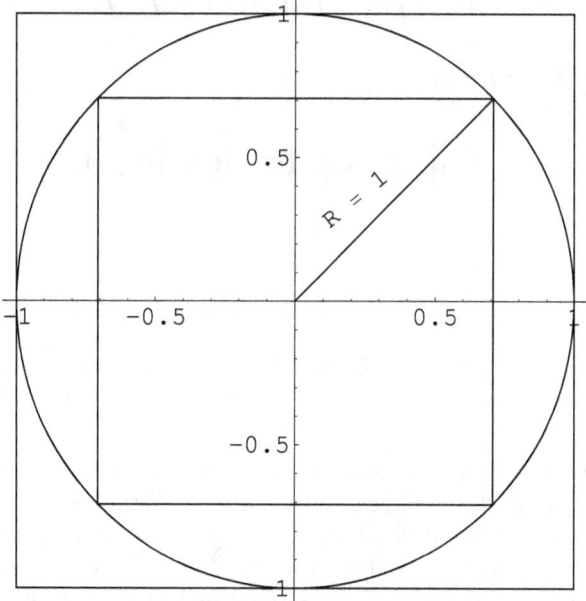

Starting with a square ($N = 4$) we obtain $2 < \pi < 4$, next for octagon $N = 8$ $2\sqrt{2} < \pi < 8(\sqrt{2} - 1)$, or $2.83 < \pi < 3.31$, and so forth.

In general, we start with $N = 4$ and at each iteration we double the number of sides: $N \to 2N$. The angles α_n of the corresponding triangles are diminished by two. The quantity $s = \sin(\alpha)$ satisfies the equation

$$4s^4 - 4s^2 + y^2 = 0 \qquad (10.2)$$

where $y = \sin(2\alpha)$. Therefore, after n steps for the angles α_n of inscribed triangles we obtain

$$s_{n+1} = \sqrt{\frac{1 - \sqrt{1 - s_n^2}}{2}}, \quad s_0 = 1 \qquad (10.3)$$

where $s_n = \sin(\alpha_n)$. In a similar way for $t = \tan(\alpha)$ we have the equation

$$zt^2 + 2t - z = 0 \qquad (10.4)$$

where $z = \tan(2\alpha)$, and for circumscribed polygons we obtain

$$t_{n+1} = \frac{\sqrt{1 + t_n^2} - 1}{t_n}, \quad t_0 = 1 \qquad (10.5)$$

Using those formulae for inscribed polygons, I wrote this program:

```
piInscribed[s_] :=
 (* Compute an approximation of pi after s steps
    using areas of inscribed polygons. *)
  Module[{n=2., sn=1., i},
      For[i=1, i <= s, i++,
          sn = Sqrt[ (1-Sqrt[1-sn*sn])/2];
          n *= 2];
      sn * n]
```

The behavior of the function piInscribed was very interesting. For small but increasing values of s it gave values closer and closer to π (from below). Then the approximations of π became worse, and finally jumped to 0 and stayed there. This was the way my students and I 'proved' that $\pi = 0$.

Looking carefully at this function it is pretty clear what happened: s_n diminishes quite rapidly, so $1 - s_n^2$ becomes a clean number one.

$1 - \sqrt{(1 - s_n^2)}$ becomes zero, and when we multiply by zero, we still get zero. It does not mean that $\pi = 0$, but that the formulas (10.3) and (10.5) are not useful for calculations.

We have shown that when we subtract two $\mathcal{R}$ numbers that are close together we can lose accuracy. For subtraction in a computer, $1 - \sqrt{1 - \epsilon}$ is very bad. Since

$$\frac{1 - \sqrt{1 - s^2}}{2} = \frac{s^2}{2(1 + \sqrt{1 - s^2})}, \quad \frac{\sqrt{1 + t^2} - 1}{t} = \frac{t}{1 + \sqrt{1 + t^2}} \quad (10.6)$$

we can eliminate the bad subtraction by using division instead.

(We obtain equation (10.6) from the formula $a - b = (a^2 - b^2)/(a+b)$.)

Notice also that multiplication of a very large number by a very small positive one such as $N_i * \epsilon_i$ is dangerous because it can cause overflow or underflow.

In our case it is better to change the variable and use $\pi_i = 2^i * s_i$ which will be nearly constant around 3. Using (10.6) and eliminating s_i, we obtain the formula:

$$\pi_{i+1} = \pi_i \sqrt{\frac{2}{1 + \sqrt{1 - \pi_i^2 4^{-i-1}}}}, \quad \pi_0 = 2 \quad (10.7)$$

for $i = 0, 1, \ldots$. As i increases, the expression under the square root goes to 1 and when it reaches 1, the next π_{i+1} will be equal to π_i.

We can trace all our troubles to sloppy solutions of the quadratic equations (10.2) and (10.4). For a detailed discussion of solutions to a general quadratic equation, see Appendix D.

We used the formulae for areas of polygons inscribed into the circle because of the very simple boundary conditions $s_0 = t_0 = 1$ in formula (10.1). Let us return to the computation of π based on circumferences (as was done by Archimedes). Since the circumference of the polygon with N sides inscribed in the unit circle ($R = 1$) is $o_N = 2N \sin(\pi/N)$, and the circumscribed one is $O_N = 2N \tan(\pi/N)$, we have

$$N \sin(\pi/N) < \pi < N \tan(\pi/N) \quad (10.8)$$

Using two terms of the Taylor series for large N (good to $O(N^{-4})$), we obtain

$$N\sin(\pi/N) \approx \pi - \frac{\pi^3}{6N^2} + \cdots, \quad N\tan(\pi/N) \approx \pi + \frac{\pi^3}{3N^2} + \cdots \tag{10.9}$$

Eliminating the term $1/N^3$ we obtain Snellius' (1580-1626) formula

$$\frac{2}{3}N\sin(\pi/N) + \frac{1}{3}N\tan(\pi/N) \to \pi \tag{10.10}$$

which gives $\pi \approx 2 + \frac{2}{3}\sqrt{3}$ for $N = 6$ (a hexagon), and $\pi \approx 3.141592...$ for $N = 96$.

In 1800, F. Pfaff proved that starting with $a_0 = 2\sqrt{3}$ and $b_0 = 3$, where a_n and b_n are circumferences of inscribed and circumscribed polygons with $6 \cdot 2^n$ sides respectively, gives

$$a_{n+1} = \frac{2a_n b_n}{a_n + b_n}, \quad b_{n+1} = \sqrt{a_{n+1} b_n} \tag{10.11}$$

and

$$b_n < b_{n+1} < a_{n+1} < a_n, \quad \pi = \lim_{n \to \infty} a_n = \lim_{n \to \infty} b_n \tag{10.12}$$

Substituting in (10.11) $\alpha_n = 1/a_n$ and $\beta_n = 1/b_n$, we obtain these beautiful formulae for the calculation of π:

$$\alpha_{n+1} = \frac{\alpha_n + \beta_n}{2}, \quad \beta_{n+1} = \sqrt{\alpha_{n+1}\beta_n}, \quad \alpha_0 = \sqrt{3}/2, \quad \beta_0 = 1/3 \tag{10.13}$$

The Taylor series methodology yielded more accurate results for the computation of π than the Archimedean methods. A most important formula was discovered by J. Gregory (1638–1675). He found that the area under the curve $1/(1 + x^2)$ is $\arctan(x)$, or

$$\int_0^x \frac{dz}{1+z^2} = \arctan(x) \tag{10.14}$$

He then obtained the series

$$\arctan(x) = x - \frac{x^3}{3} + \frac{x^5}{5} + \cdots + (-1)^n \frac{x^{2n+1}}{2n+1} + \cdots \tag{10.15}$$

which to this day bears Gregory's name.

The formula (10.15) can be obtained by integrating term by term the geometric series

$$\frac{1}{1+x^2} = 1 - x^2 + x^4 + \cdots + (-1)^n x^{2n} + \cdots \qquad (10.16)$$

Then substitute $x = 1$ in (10.15). Since $\arctan(1) = \pi/4$ this yields

$$\pi = 4\left(1 - \frac{1}{3} + \frac{1}{5} + \cdots + \frac{(-1)^n}{2n+1} + \cdots\right) \qquad (10.17)$$

However, the convergence of this series is too slow for any practical computation.

The astronomer A. Sharp (1651–1792) substituted $x = 1/\sqrt{3}$ in (10.15), which gave him

$$\pi = 2\sqrt{3}\left(1 - \frac{1}{3\cdot 3} \cdots + \frac{(-1)^n}{(2n+1)3^n} + \cdots\right) \qquad (10.18)$$

and using this series he obtained 72 decimal places.

Newton used his binomial theorem to obtain

$$\arcsin(x) = \int_0^x \frac{dz}{\sqrt{1-z^2}}$$
$$= \int_0^x dz \left(1 + \frac{z^2}{2} + \frac{1\cdot 3}{2\cdot 4}z^2 + \cdots + (-1)^n \binom{-1/2}{n} z^{2n} + \cdots \right) \qquad (10.19)$$

Substituting $x = 1/2$ in (10.19) makes $\arcsin(1/2) = \pi/6$. This yields

$$\pi = 6\left(\frac{1}{2} + \frac{1}{2\cdot 3 \cdot 2^3} + \frac{1\cdot 3}{2\cdot 4 \cdot 5 \cdot 2^5} + \cdots\right) \qquad (10.20)$$

In 1706 J. Machin, a professor of astronomy in London, proved that

$$\pi/4 = \arctan(1) = 4\arctan(1/5) - \arctan(1/239) \qquad (10.21)$$

The proof is based on the formula

$$\arctan(x) + \arctan(y) = \arctan\left(\frac{x+y}{1-x\cdot y}\right) \qquad (10.22)$$

which follows from
$$\tan(\alpha + \beta) = \frac{\tan(\alpha) + \tan(\beta)}{1 - \tan(\alpha) \cdot \tan(\beta)} \tag{10.23}$$

Substituting $x = y$ in (10.22) yields
$$2\arctan(x) = \arctan\left(\frac{2x}{1 - x^2}\right) \tag{10.24}$$

Applying (10.24) twice with $x = 1/5$ yields
$$4\arctan\left(\frac{1}{5}\right) = 2\arctan\left(\frac{5}{12}\right) = \arctan\left(\frac{210}{119}\right) \tag{10.25}$$

Therefore, it follows from (10.22) that
$$\begin{aligned}\arctan(1) - 4\arctan(1/5) &= \arctan(1) - \arctan\left(\frac{210}{119}\right) \\ &= -\arctan\left(\frac{1}{239}\right)\end{aligned} \tag{10.26}$$

This proves (10.21).

From (10.21), Machin calculated π to 100 places in 1706. Machin's method was used by Euler, who developed a formula
$$\frac{\pi}{4} = 5\arctan\left(\frac{1}{7}\right) + 2\arctan\left(\frac{3}{79}\right) \tag{10.27}$$

which can be proved similarly to (10.21),
Evaluating (10.27) using the formula
$$x \cdot \arctan(x) = y\left(1 + \frac{2}{3}y + \frac{2 \cdot 4}{3 \cdot 5}y^2 + \cdots\right) \tag{10.28}$$

where $y = x^2/(1 + x^2)$, Euler calculated π to 20 places in one hour.

The series (10.28) converges faster than Gregory's formula for small $|x|$. To prove (10.28) notice that
$$\arctan(x) = \arcsin\left(\sqrt{\frac{x^2}{1 + x^2}}\right) \tag{10.29}$$

and

$$\frac{\arcsin(x)}{\sqrt{1-x^2}} = x + \frac{2}{3}x^3 + \frac{2\cdot 4}{3\cdot 5}x^5 + \cdots + \frac{(2n)!!}{(2n+1)!!}x^{2n+1} + \cdots \quad (10.30)$$

Combining (10.29) and (10.30) yields (10.28).

The Taylor series (10.30) for $\arcsin(x)/\sqrt{1-x^2}$ was obtained in the previous chapter, and can be used for the decimal computation of π. The formula can be obtained from the identity

$$\frac{\pi}{4} = 2\arctan(\frac{1}{3}) + \arctan(\frac{1}{7}) \quad (10.31)$$

which can be proved using (10.22).

The series expansion (10.28) yields

$$\frac{\pi}{4} = \sum_{n=0}^{\infty} \frac{(2n)!!}{(2n+1)!!}\left(6\cdot 10^{-n-1} + 7\cdot 50^{-n-1}\right) \quad (10.32)$$

The computation of π by power series changed its character since the famous work of D. Bailey, P. Borwein, and S. Plouffe who in 1996 published the series

$$\pi = \sum_{k=0}^{\infty} \frac{1}{16^k}\left(\frac{4}{8k+1} - \frac{2}{8k+4} - \frac{1}{8k+5} - \frac{1}{8k+6}\right) \quad (10.33)$$

This series has a remarkable property that it allows the computation of an arbitrary hexadecimal digit of π without knowledge of the previous digits, therefore allowing parallel computation.

The proof of the formula (10.33) is simple. We have

$$\int_0^{1/\sqrt{2}} \frac{x^k}{1-x^8}\,dx = \int_0^{1/\sqrt{2}} \sum_{i=0}^{\infty} x^{k+8i}\,dx = 2^{-(k+1)/2}\sum_{i=0}^{\infty}\frac{1}{16^i(8i+k+1)} \quad (10.34)$$

The integrals in this formula are elementary, and involve log and arctan functions. Here is a *Mathematica* function.

```
sm[k_]:=
   2^(k+1)/2 Integrate[x^k/(1-x^8),
                  {x,0,1/Sqrt[2]}];
   pp=4sm[0]-2sm[3]-sm[4]-sm[5]//Simplify
```

Mathematica prints out π.

D. Bailey et. al. have shown that the digits in the base b expansion of S are given by

$$S = \sum_{k=0}^{\infty} \frac{p(k)}{b^{ck}q(k)} \qquad (10.35)$$

where p and q are polynomials with integer coefficients, and b and c are positive integers. Beginning at position n can be obtained from the fractional part of b^n. We have

$$b^n S \bmod 1 = \sum_{k=0}^{\lfloor n/c \rfloor} \frac{b^{n-ck}q(k) \bmod p(k)}{p(k)} \bmod 1 \qquad (10.36)$$

$$+ \sum_{k=\lfloor n/c \rfloor+1}^{\infty} \frac{b^{n-ck}p(k)}{q(k)} \bmod 1$$

We should compute only enough terms of the second sum using floating point arithmetic to ensure the truncation error is smaller than $1/b$.

The series (10.32) discovered by the author can be transformed into

$$\frac{\pi}{4} = \sum_{n=0}^{\infty} 100^{-n-1} \left(24 \cdot (11n+8) \frac{(4n)!!}{(4n+3)!!} + 14 \cdot 2^n \frac{(2n)!!}{(2n+1)!!} \right) \qquad (10.37)$$

which gains 2 digits with each term and is in the form of (10.35).

The next series involving π which we present here is the famous Ramanujan's sum:

$$\frac{1}{\pi} = \frac{\sqrt{8}}{9801} \sum_{n=0}^{\infty} \frac{(4n)![1103 + 26390 \cdot n]}{(n!)^4 \, 396^{4n}} \qquad (10.38)$$

The proof of this sum is beyond the scope of this book, but the speed of convergence is astounding: each term adds 8 digits.

The methods which lead to very fast algorithms are iterative and based on the theory of elliptic functions. We will not deal with them here.

Recent computations of π were done in 2002 by T. Kanada of University of Tokyo. He used Machin's type of series invented in 1982 by K. Takano:

$$\pi = 48 \cdot \arctan \frac{1}{49} + 128 \cdot \arctan \frac{1}{57} - 20 \cdot \arctan \frac{1}{239} \\ + 48 \cdot \arctan \frac{1}{110443} \quad (10.39)$$

Using this formula he obtained 1.4 trillion digits. He checked them with the formula invented by F. Störmer in 1886:

$$\pi = 176 \cdot \arctan \frac{1}{57} + 28 \cdot \arctan \frac{1}{239} - 48 \cdot \arctan \frac{1}{682} \\ + 96 \cdot \arctan \frac{1}{12943} \quad (10.40)$$

We can find many more formulae involving π in the beautiful book π *Unleashed* (J. Arndt and C. Haenkel, Springer, 2001).

Chapter 11

In the footsteps of Euclid, who over 2000 years ago knew the beauty of proof and of number theory

(Number theory programs)

Number theory is one of the oldest and prettiest parts of mathematics. One of the fundamental problems of number theory is the study of divisibility of numbers. Let d and n be natural numbers. A number d divides n if there is a natural number x such that $x \cdot d = n$. In other words, division of n by d leaves no remainder. We can write

Illustration 11.1 *divides.*

```
divides[d_?Positive,n_?Positive] := Mod[n,d] == 0;
```

A prime number is a natural number greater than 1 that has no divisors other than 1 and itself. The first seven prime numbers are

$$2, 3, 5, 7, 11, 13, 17, ...$$

Euclid knew many fundamental facts about prime numbers. His proof of the existence of an infinite number of primes is as pretty today as it was more than 2000 years ago. Here it is:

Suppose there are only n prime numbers $p_1, p_2, ..., p_n$. Consider $m = 1 + \prod_{i=1}^{n} p_i$. What are divisors of m? It is not divisible by any p_i, since division by p_i leaves a remainder of 1, therefore if m is prime, it is a new prime not on the given list, and if m is not prime it must be divisible by some new prime. This contradicts the assumption that $p_1, p_2, ..., p_n$ was the full list of primes.

To write a simple prime number test, we introduce a function LD for Least Divisor. Let $n > 1$ be a natural number, then $L = LD(n)$ is the least natural number greater than one that divides n. Such a number always exists and is prime. L always exists since the number $L = n$ is greater than 1 and is a divisor. L must be prime. If it were not prime, then there would exist a smaller divisor, which is impossible.

It is easy to write a more general program $LDS(k, n)$: the least divisor of n starting with k. We obtain our $LD(n) = LDS(2, n)$. We search for k up to $k \leq \sqrt{n}$, or $k * k \leq n$, and increment k.

Illustration 11.2 *Least divisor ld[n].*

```
ld[n_]:= lds[2,n]; (* Least divisor of n *)

lds[k_,n_]:= (* Least divisor of n starting with k *)
  If[divides[k,n], k,
     If[k * k > n, n, lds[k + 1, n] ]
  ]

ld[25] 5
```

Now we can implement a test for n being prime number:

Illustration 11.3 *Prime number test.*

```
prim[n_]:= ld[n] == n
```

Now that we know how to decide whether n is a prime number, let us consider the prime factors of any $m > 1$. A list of prime numbers $p_1, \ldots, p_i$ with the property that the product

$$p_1 \cdot p_2 \cdots p_i = m$$

is called the *prime factorization* of m. For instance the prime factorization of 12 is {2, 2, 3}, since 12 = 2 · 2 · 3, and 2 and 3 are prime numbers. Here is a program that implements prime factorization:

Illustration 11.4 *Prime factorization of n.*

```
factors[n_]:=
  Module[{p, res, m = n},
         For[res={}, m > 1,,
             p = ld[m];
             m /= p;
             res = AppendTo[res, p]];
         res];

factors[12]  {2, 2, 3}
```

The theorem stating that every natural $n > 1$ has a unique prime factorization is called the **fundamental theorem of arithmetic**, and was known in antiquity. The program written above proves that a factorization exists, since this program always produces an answer. To prove the uniqueness of this answer, we need the concept of "the greatest common divisor", or $GCD(m, n)$, of two natural numbers.

$d = GCD(m, n)$ is a number d that divides both m and n and, for all natural numbers d' that divide both m and n, it holds that d' divides d. For example $GCD(48, 30) = 6$, because 6 divides 48 and 30, and every other common divisor of 48 and 30 divides 6 (the other common divisors are 1,2,3). Clearly d must be unique, for if d' also qualifies, then because d' divides m and n, it follows from the definition of d that d' divides d, and it must be that $d = d'$.

The $GCD(m, n)$ can be found using the prime factorization of m and n, by choosing the minima of powers of the prime factors. For instance $48 = 2^4 \cdot 3$, $30 = 2 \cdot 3 \cdot 5$, therefore $GCD(48, 30) = 2 \cdot 3 = 6$.

But there is an easier way to find $GCD(m, n)$. It is Euclid's famous algorithm. Euclid's algorithm has many versions in different sciences: in number theory, geometry, in the theory of polynomials, and other parts of mathematics.

In geometry GCD deals with measuring the length of two line segments. Given two segments A and B, find their *greatest common measure*, the largest segment C that covers both segments A and B completely.

Euclid proposed the following solution: Take the shorter segment; suppose it is B. (If it is A, change the names.). Then
1. Put the segment B on the segment A as many times as possible. There are 2 possibilities:
2. The end of the segment A coincides with the end of the segment B. If so B is the greatest common measure, since it a common measure and it cannot be bigger than one of the segments.
3. If there is some segment left over, call it C, and repeat steps 1–2 with the segments B and C.

Example: let the lengths of segments A and B be 24 and 15 units, respectively. Then using Euclid's method we get the following triples A, B, C:

$$(24, 15, 9), \ (15, 9, 6), \ (9, 6, 3), \ (6, 3, 0)$$

After 4 repetitions of steps 1, 2, and 3, we obtain the greatest common measure, the segment whose length is 3.

One trouble with this reasoning appeared in Pythagorean times when mathematicians tried to find the greatest common measure of a side of the square and its diagonal without success. The diagonal of the unit square is equal to $\sqrt{2}$, and 1 and $\sqrt{2}$ have no common measure! In our case for the natural numbers, this trouble cannot happen, since the number 1 is a common measure for any natural number. In this case, the greatest common measure is called the greatest common divisor $GCD(m, n)$, and using the function Mod[m,n] giving remainder of the division of m by n, we can formulate Euclid's algorithm:

Illustration 11.5 *Euclid's Algorithm as a function.*

```
gcdf[m_Integer?Positive, n_Integer?Positive]:=
(* Compute GCD(m,n) for natural numbers n, n *)
  If[n == 0, m, gcdf[n, Mod[m , n]]];

gcdf[24,15]   3
```

We can write also this function in an imperative style:

Illustration 11.6 *Imperative Euclid's Algorithm.*

```
gcdi[ mm_Integer, nn_Integer]:=
(* Compute GCD(m, n) for natural m, n, imperatively *)
  Module[{r, m = mm, n = nn},
        (* Validate m, n *)
        If[m <= 0 || n <= 0, 0,
            For[r=1, r != 0, m = n; n = r,
                r = Mod[m, n] ];
            m]];

gcdi[24, 15] 3
```

We have a small trouble here: if m is a parameter, say 24, then the operation $m = n$ makes no sense, because the left hand side of the replacement has to be a name, not a constant. Therefore, we introduce two local variables. Some languages treat names of parameters inside the function as local variables. *Mathematica* does not.

We do not worry about the ordering of the numbers m and n, since in if $m < n$, the loop is repeated once more, putting them in increasing order. The loop in the *gcdi* function terminates because computed remainders form a strictly decreasing sequence and these remainders are all nonnegative, so there has to come a moment when we obtain the number zero.

This function computes GCD because a triple $(m, n, r = m - qn)$ has the same divisors, therefore also the largest one. The function $GCD(m, n)$ works in logarithmic time. This follows from Lame's theorem, which says that

Theorem 11.1
If $n' = min(m, n)$, and the loop in $GCD(m, n)$ uses k steps, then $n' \geq fib(k)$, where $fib(k)$ is the Fibonacci number f_k.

In our example for the pair 24, 15 we have:

$$15 \geq fib(4) = 3$$

As you remember, we proved that

$$Fibonacci(k) = Round(\phi^k/\sqrt{5})$$

where $\phi \approx 1.62$, therefore we can compute k from

$$n' \geq Round(\phi^k/\sqrt{5})$$

and the number of steps is logarithmic.

We will sketch the proof of Lame's theorem:

Let k be a number of repetitions of the loop in the gcdi program, and let (m_k, n_k), (m_{k-1}, n_{k-1}), and (m_{k-2}, n_{k-2}) be the values of the last three pairs (m, n). We have $n_k \leq n_{k-1} + n_{k-2}$, since at each step there is a $q \geq 1$ such that $m_i = qn_i + n_{i-1} \geq n_i + n_{i-1}$, and in the previous step we had $n_{i+1} = m_i$. To finish the proof it is enough to notice that $n_1 \geq 1$, and that we used the equation $f_{n+1} = f_n + f_{n-1}$, where $f_n = Fibonacci(n)$.

There exists a version of Euclid's algorithm which is even closer to the *gcdi* given above:

Illustration 11.7 *Euclid's Algorithm without division.*

```
gcdw[mm_Integer, nn_Integer]:=
(* Compute GCD(m, n) without division *)
  Module[{n = nn, m = mm},
      If[m <= 0 || n <= 0, 0,
          While[n != m,
              If[n > m, n -= m];
              If[n < m, m -= n];
          ];
      ]
    m];

gcdw[24, 15]  3
```

For our example we obtain $(24, 15) \rightarrow (9, 6) \rightarrow (3, 3)$ very quickly. Unfortunately, in the case where the numbers m and n are large and close together, the loop is executed many times. Attempts to improve this simple and pretty algorithm can go in many directions. For instance, if one of the numbers becomes 1, then the whole GCD is 1, but checking many special cases usually does not improve an algorithm.

A more promising approach is to check systematically in the loop:

Illustration 11.8 *Another Euclid's algorithm without division.*

1. Both even: $GCD(2m, 2n) = 2\,GCD(m, n)$

2. Odd(n): $GCD(2m, n) = GCD(m, n)$

3. $m > n$: $GCD(m - n, n) = GCD(m, n)$

4. $GCD(n, m) = GCD(m, n)$

In some applications we need the following generalization of the Euclidean Algorithm: find $GCD(m, n)$ and two integers x, y such that

$$m \cdot x + n \cdot y = GCD(m, n) \quad (11.1)$$

The solution to this problem has many applications, for instance in the theory of linear Diophantine equations, and in cryptography.

There are many possible methods of computing the numbers x, y and GCD. We will sketch some of them, leaving the actual programs as exercises.

Let us start with a pair $(24, 15)$. From the Euclidean algorithm, we have

$$24 = 15 \cdot 1 + 9; \ 15 = 9 \cdot 1 + 6; \ 9 = 6 \cdot 1 + 3; \ 6 = 3 \cdot 2 + 0 \quad (11.2)$$

Now eliminate the remainders r_i, starting from the end in equation (11.2) to obtain

$$3 = 9 - 6 \cdot 1; \ 6 = 15 - 9 \cdot 1; \ 9 = 24 - 15 \cdot 1 \quad (11.3)$$

Substituting the remainders yields

$$3 = 9 + (-1)(15 - 9) = 2 \cdot 9 - 15 = 2(24 - 15) - 15 = 2 \cdot 24 - 3 \cdot 15 \quad (11.4)$$

In this way we obtain a pair of numbers $x = 2$, $y = -3$ such that equation (11.1) is satisfied.

In general, let $a_0 = m$, $a_1 = n$, and let a_k be the last nonzero remainder. Using the Euclidean algorithm, we obtain

$$a_0 = a_1 q_1 + a_2, \quad \text{where} \quad 0 \leq a_2 < a_1$$
$$a_1 = a_2 q_2 + a_3, \quad \text{where} \quad 0 \leq a_3 < a_2$$
$$\vdots$$
$$a_i = a_{i+1} q_{i+1} + a_{i+2}, \quad \text{where} \quad 0 \leq a_{i+2} < a_{i+1} \quad (11.5)$$
$$\vdots$$
$$a_{k-2} = a_{k-1} q_{k-1} + a_k, \quad \text{where} \quad 0 \leq a_k < a_{k-1}$$
$$a_{k-1} = a_k q_k + 0$$

Studying these equations (11.5), we have many possible programming approaches:

1. Write a_i and r_i until $i = k$ in an array, and use this array to compute the numbers x, y in a way similar to what we did in the last example.

2. Notice that only the r_i are needed, since the equation

$$a_i = a_{i+1}q_{i+1} + a_{i+2} \qquad (11.6)$$

is linear and $a_{k+1} = 0$. Setting $b_k = 0$, $b_{k-1} = 1$, we obtain for the computation of b_i the equations

$$b_{i-1} = b_i q_i + b_{i+1}, \quad \text{where} \quad i = k-2, \cdots, 2, 1 \qquad (11.7)$$

therefore b_i depends only on the numbers q_i.

The above method gives $b_0 = |y|$, $b_1 = |x|$.

We obtain the correct signs by changing them in each step, obtaining for an even k pair $b_0 = -y$, $b_1 = x$, and for an odd k $b_0 = y$, $b_1 = -x$. For example, for the pair (24, 15) we obtain

i	0	1	2	3	4
a_i	24	15	9	6	3
q_i	-	1	1	1	1
b_i	3	2	1	1	0

Since the last k such that $a_k \neq 0$ is $k = 4$, we have $x = 2$, $y = -3$, giving

$$2 \cdot 24 - 3 \cdot 15 = GCD(24, 15) = 3 \qquad (11.8)$$

and for the pair (246, 194)

i	0	1	2	3	4	5	6	7
a_i	246	194	52	38	14	10	4	2
q_i	-	1	3	1	2	1	2	2
b_i	52	41	11	8	3	2	1	0

we have $k = 7$, therefore

$$-41 \cdot 246 + 52 \cdot 194 = GCD(246, 192) = 2 \qquad (11.9)$$

Another possibility is to use recursion and generalize the function *gcdf*.

Let us return to the problem of finding prime numbers. One of the oldest and fastest methods is the Sieve of Eratosthenes, invented by Eratosthenes about 2500 years ago.

To find all primes less than *maxPrime*, we delete successively numbers $n > 2$ divisible by 2, $n > 3$ divisible by 3, $n > 5$ divisible by 5, and so on. The remaining numbers have to be prime. We also notice that we have to check the numbers only to $\sqrt{maxPrime}$ (why?). A first attempt may look as follows:

Illustration 11.9 *Prime numbers computed by the Sieve of Eratosthenes.*

```
sieveEratosthenes[n_Integer]:=
(* Compute prime numbers sieve[[i]] < n *)
  Module[
    {i, count, number, sieve, maxPrime = 100000},
    If[n < 2 || n > maxPrime, count = 0,
        (* sieve: successive natural numbers *)
        sieve = Table[i,{i, n}];
        For[number = 2, number * number < n, number++,
            For[i = 2*number, i <= n, i += number,
                (* Mark non-primes with 1 *)
                sieve[[i]] = 1
            ]
        ];
        (* Remove 1s, which are not prime *)
        sieve = DeleteCases[sieve, 1]
    ];
    (* Uncomment Print to see the list of primes *)
    (* Print[sieve]; *)
    count = Length[sieve]
  ]

sieveEratosthenes[30]    10
```

Here *Length*[*list*] gives the length of the list, and *DeleteCases*[*list*, *n*] deletes every number *n* from the list.

Another program tests the divisibility of *n* by successive natural numbers starting with $n = 2$ instead of deleting numbers. Here is that program:

Illustration 11.10 *Compute primes testing for divisibility.*

```
nPrimes[nmax_Integer]:=
(* Compute nmax primes *)
  Module[{lim = 1, n = 1, pr, k, j, t},
         p[[1]] = 2;
         For[j = 1, j <= nmax, j++,
             pr = False;
             While[!pr,
                   n += 2;
                   t = p[[lim]];
                   If[t*t <= n, lim++];
                   k=2;
                   pr = True;
                   While[pr && (k < lim),
                         k++;
                         pr=(Mod[n , p[[k]]] != 0);
                         k++]
                  ];
             p[[j]] = n]];

nn=10; p = Table[0,{nn}]; nPrimes[nn];
p   {2, 3, 5, 7, 11, 13, 17, 19, 23, 29}
```

One very interesting test for prime numbers is based on the "Fermat's little theorem."

Theorem 11.2 *If n is a prime number and a is a natural number $a < n$, then $a^n \bmod n = a$.*

This kind of test is called a probabilistic test because if n does not satisfy Fermat's test, we can be sure that n is not prime; on the other hand, if n satisfies the test we are not sure, since there are numbers that only pretend to be prime. These numbers are called Carmichael numbers, and they are quite sparse. The first three Carmichael numbers are 561, 1105 and 1729.

To write a program using the Fermat Test, we need a fast computation of $a^n \bmod m$. One possibility is to use a modification of our program to compute a^n in logarithmic time (program 5.5):

Illustration 11.11 *Fast computation of powers mod m*

```
modPow[x_Integer?Positive, n_Integer?Positive,
      m_Integer?Positive]:=
(* Compute x^n mod m. *)
  Module[{k, pr = 1, db},
         db = Mod[x,m];
         For[k = n, k > 1, k = Quotient[k,2],
             If[OddQ[k],
                pr = Mod[pr*db, m]]; (* collect *)
                db = Mod[db * db, m]]; (* double *)
         Mod[db * pr, m]];

modPow[2,5,10]   2
```

The set of numbers which do not satisfy a probabilistic test and hide a few composite numbers we call *pseudo-primes*. The numbers that pass the Fermat test are pseudo-primes.

A different set of pseudo-primes use the Perrin numbers (P_n) test explained below. P_n are similar to Fibonacci numbers. They are computed from the following formulae:

$$P_n = P_{n-2} + P_{n-3}, \quad P_0 = 3, \quad P_1 = 0, \quad P_2 = 2 \qquad (11.10)$$

Below we write the first 14 Perrin numbers:

n	0	1	2	3	4	5	6	7	8	9	10	11	12	13
P_n	3	0	2	3	2	5	5	7	10	12	17	22	29	39
P_n mod n	-	0	0	0	2	0	5	0	2	3	7	0	5	0

Notice here the remarkable fact that if n is a prime number, then (as Lucas has proved) P_n mod $n = 0$. The important question is: does there exist an n such that P_n mod $n = 0$, but n is not prime? Unfortunately, there does. The smallest such number is $521^2 = 271441$. However, such numbers are very sparse.

Looking for a computation of Perrin numbers in logarithmic time, we assume $P_n = A\lambda^n$ and obtain for λ a third degree equation

$$\lambda^3 - \lambda - 1 = 0 \qquad (11.11)$$

that has one real root

$$\lambda_1 = \sqrt[3]{\left(9+\sqrt{69}\right)/18} + \sqrt[3]{\left(9-\sqrt{69}\right)/18} \approx 1.32471\ 79572\ 44746 \tag{11.12}$$

and two complex roots

$$\lambda_2 \approx -0.662359 - 0.56228i, \quad \lambda_3 \approx -0.662359 + 0.56228i \tag{11.13}$$

The numbers λ_2 and λ_3 can be written in terms of λ_1, because they are the roots of the equation

$$x^2 + \lambda_1 x + 1/\lambda_1 = 0 \tag{11.14}$$

This equation follows from Vieta's formulas for equation (11.11):

$$\lambda_1 + \lambda_2 + \lambda_3 = 0, \quad \lambda_1\lambda_2\lambda_3 = 1 \tag{11.15}$$

The Perrin numbers P_n can be expressed by

$$P_n = A\lambda_1^n + B\lambda_2^n + C\lambda_3^n \tag{11.16}$$

for $n = 0, 1, \ldots$.

If we use the conditions $P_0 = 3$, $P_1 = 0$, $P_2 = 2$, we obtain a system of 3 equations for the constants A, B, and C. Solving the system we get

$$A = B = C = 1$$

so we have

$$P_n = \lambda_1^n + \lambda_2^n + \lambda_3^n \tag{11.17}$$

We notice that $\lim_{n\to\infty} \lambda_2^n = \lim_{n\to\infty} \lambda_3^n = 0$, and we have

$$P_n \approx \lambda_1^n \approx p_n = (1.32472\ldots)^n. \tag{11.18}$$

The following table illustrates the relation between P_n and p_n.

n	0	1	2	3	4	5	6	7	8	9	10	11	12
P_n	3	0	2	3	2	5	5	7	10	12	17	22	29
p_n	1.0	1.3	1.8	2.3	3.1	4.1	5.4	7.2	9.5	12.6	16.6	22.0	29.2

We can see that
$$P_n = Round(\lambda_1^n) \tag{11.19}$$
For a quick computation of Perrin numbers, we introduce the Padavan numbers Q_n. They satisfy the same recurrence relation, but with different boundary conditions:
$$Q_n = Q_{n-2} + Q_{n-3}, \quad Q_0 = 0, \quad Q_1 = 1, \quad Q_2 = 1 \tag{11.20}$$
We can prove that
$$Q[n] = \begin{pmatrix} 0 & 1 & 1 \\ 1 & 0 & 0 \\ 0 & 1 & 0 \end{pmatrix}^n = \begin{pmatrix} Q_{n-1} & Q_n & Q_{n-2} \\ Q_{n-2} & Q_{n-1} & Q_{n-3} \\ Q_{n-3} & Q_{n-2} & Q_{n-4} \end{pmatrix} \tag{11.21}$$
To prove it by induction we have to add a few Padavan numbers with negative indexes:
$$Q_{-3} = Q_{-2} = 0, \quad Q_{-1} = 1 \tag{11.22}$$
Using these numbers the proof is immediate.

Since $Q[n] \cdot Q[m] = Q[n+m]$, we have
$$\begin{aligned} Q_{2n} &= Q_{n-2}^2 + 2Q_n Q_{n-1} \\ Q_{2n+1} &= Q_n^2 + Q_{n-1}^2 + 2Q_{n-1}Q_{n-2} \\ Q_{2n+2} &= Q_{n-1}^2 + 2Q_n(Q_{n-1} + Q_{n-2}) \end{aligned} \tag{11.23}$$
and
$$\begin{aligned} Q_{n+m} &= Q_{n-1}Q_m + Q_n Q_{m-1} + Q_{n-2}Q_{m-2} \\ Q_{n+m-1} &= Q_{n-1}Q_{m-1} + Q_n Q_{m-2} + Q_{n-2}(Q_m - Q_{m-2}) \\ Q_{n+m-2} &= Q_{n-1}Q_{m-1} + Q_{n-1}Q_{m-2} + (Q_n - Q_{n-2})(Q_m - Q_{m-2}) \end{aligned} \tag{11.24}$$
Finally we can get to the Perrin numbers P_n, using the fact that P_n and Q_n satisfy the same recurrence equation. We have
$$P_n = 3Q_{n-1} - Q_{n-3} \tag{11.25}$$
which we get by computing the constants A, B, and C in the equation
$$P_n = A \cdot Q_{n-1} + B \cdot Q_{n-2} + C \cdot Q_{n-3} \tag{11.26}$$
Collecting all of this information we obtain the following program:

Illustration 11.12 *Fast computation of Perrin numbers.*

```
perrin[n_Integer?Positive]:=
(* Fast computation of the Perrin numbers. *)
  Module[
    {k=n+1, p0=1, p1=0, p2=0, d0=1, d1=0, d2=1, t1, t2},
    While[k != 1,
      If[OddQ[k],
        t2 = d0*p0 + d1*p2 + d2*p1;
        t1 = d0(p2-p0)+ d1* p1 + d2*p0;
        p0 = d0*p1 + d1*p0 + (p2-p0)*(d2-d0);
        p1 = t1; p2 = t2]; (* folding *)
      k = Quotient[k,2];
      t2 = d0*d0 + 2*d1*d2;
      t1 = d1*d1 -d0*d0 + 2*d0*d2;
      d0 = (d2-d0)*(d2-d0) + 2*d0*d1;
      d1 = t1;
      d2 = t2]; (* copying *)
    3*(d0*p0 + d1*p2 + d2*p1)
    -(d0*p1 + d1*p0 + (p2-p0)*(d2-d0))];

perrin[13]   39
```

I leave the computation of the function P_n mod n as an exercise.

One of the most interesting applications of number theory is cryptography, or encoding and decoding of secret messages making them unreadable if they are intercepted by the wrong people. *Encryption* is the transforming of a message M into a form readable only to chosen people. One of the most popular methods of encryption is called RSA, taken from the first letters of the names of the inventors of this method: Rivest, Shamir and Adelman. It is a *public key* encryption method. A key is a function whose argument is a message M. In the RSA system each user u has two keys: public (Pu_u) and private (Pr_u). When X wants to send Y a secret message $M1 = $ 'tomorrow at one', and Y replies $M2 = $ 'tomorrow ok', then:

1. X computes $e_1 = Pu_Y(M1)$ using the public key of Y (Pu_Y). Such an e_1 is called encrypted – it is normally unreadable.

2. X sends e_1 to Y.

3. Y computes $M1 = Pr(e_1)$ using the private key Pr_Y.

4. Y encrypts $e_2 = Pu_X(M2)$ using the public key of X (Pu_X).

5. Y sends e_2 to X.

6. X computes $M2 = Pr(e_2)$ using the private key Pr_X.

The functions Pr_u and Pu_u are mutually inverse for each user u. That means that for each legal message $m \in K$ we have

$$m = Pu_u(Pr_u(m)), \quad \text{and} \quad m = Pr_u(Pu_u(m))$$

for $u = X$ and Y.

For a well constructed system the functions Pr, Pu have to be quickly computable, but their inverses practically impossible to compute. An example such function is a modular raising to an integer power.

Before we discuss how to find such functions, let us write a message sending program about a meeting between X and Y. Let the general form of encoding functions be

$$f(m, x, n) = m^x \bmod n$$

and let (n, e, d) be natural numbers. X knows only the pair (n, e), and Y knows only the pair (n, d). To encrypt m, X divides m into a sequence m_i and, for each element m_i, computes $e_i = f(m_i, e, n)$ and sends this sequence e_i to Y. Y computes $m_i = f(e_i, d, n)$. In a similar way Y divides the answer into m_i parts, encrypts $e_i = f(m_i, d, n)$, and sends e_i to X. X computes $m_i = f(e_i, e, n)$ and reads it. This process is very fast (as you remember, it is logarithmic in time).
For example let $n = 253$, $e = 7$ and $d = 63$. Here is the program:

Illustration 11.13 *Encryption example.*

```
c[m_]:= PowerMod[m,e,n];
r[m_]:= PowerMod[m,d,n];
e = 7; d = 63; n = 253;
s1 = ToCharacterCode["tomorrow at one"];
s2 = Map[c,s1];
m = FromCharacterCode[Map[r,s2]]
```

```
tomorrow at one
  a1 = ToCharacterCode["tomorrow ok"];
  a2 = Map[r,a1];
  a3 = FromCharacterCode[Map[c,a2]]
tomorrow ok
```

`PowerMod[m,x,n]` computes m^x mod n, `ToCharacterCode[m]` changes the text m into the list of integer ASCII representations of this text, and `FromCharacterCode[m]` is the inverse of `ToCharacterCode[m]`.

The function `Map[<f>,<list>]` computes the list resulting from the application of `<f>` to each element of `<list>`.

This program is not realistic for many reasons: we should not group the text into a sequence of letters, but use much bigger groups. The small number n makes the breaking of the code a trivial task. Nevertheless, the sequence of operations and use of mutually inverse functions is correct.

We leave the beautiful theory of cryptography for your individual study. Here we will show how to find the numbers e, d, and n:

1. We choose 2 prime numbers p and q. They are usually large numbers of at least 1024 bits, therefore having more than 300 decimal digits. For our toy example we chose $p = 11$, $q = 23$.

2. We compute the product $n = p \cdot q$ and the function $\phi(n) = (p-1)(q-1)$ which gives us the list of relatively prime numbers to n. Our $n = 11 \cdot 23 = 253$ and $\phi(n) = 220 = 2^2 \cdot 5 \cdot 11$.

3. We choose an arbitrary e such that $1 < e < \phi(n)$ and e is relatively prime to $\phi(n)$, in other words such that $GCD(e, \phi(n)) = 1$. We chose $e = 7$.

4. We compute d, the inverse of the number e modulo $\phi(n)$, such that $d \cdot e = 1$ mod $\phi(n)$. Such an inverse always exists and is unique. In our case we obtained $d = 63$. We check that $d \cdot e = 63 \cdot 7 = 441 = 1$ mod 220.

We notice that such an inverse can be computed using the generalized Euclidean algorithm, since we know that $GCD(e, \phi(n)) = 1$, so we look for $x = d$ such that $e \cdot x + \phi(n) \cdot y = 1$. In our case we have $7 \cdot d + 220 \cdot y = 1$.

Chapter 12

In the footsteps of Euler, *who in 1759 published an analysis of the knight's tour problem*

(Backtracking programs)

Backtracking algorithms are recursive in nature and in the most cases are slow. Nevertheless, these algorithms are a very important method for the solution of many difficult problems.

Backtracking is a method of brute force trials of different possible combinations of inputs until the last one succeeds. We make one step chosen from many possible steps in the direction of the solution. If this step is a lucky one and does not lead to a blind alley, we proceed. If not we *backtrack* and try a different path.

Let us look at the maze below:

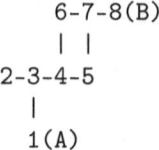

Suppose we want to find a path from the room A (numbered 1) to the room B (numbered 8). We can move only on the corridors marked '-' or '|', and we do not know where room B is until we enter it. Assuming that our rooms are numbered as in our picture, our travel might look as follows:

1. We go from room A to room 3.
2. Room 3 has two exits. Say we choose exit to room 2.
3. We have to backtrack to 3, and then we go to 4.
4. Room 4 again has two exits. Say we choose 5.
5. Room 4 was visited, so we go to 7.
6. We have two rooms not visited: 6 and 8. We choose 6, which does not lead to our goal.
7. We have to backtrack to 7. Then we go to 8 and we are in room B.

A few comments about our trip:

1. We have two kinds of objects: rooms (denoted by numbers) and corridors from-to denoted by a pair of numbers. The corridors are symmetric therefore pairs $[n, m]$ and $[m, n]$ denote the same corridor.
2. When we had a choice we started from the left hand side.
3. The corridors are rectangular, therefore in each room there may be at most four exits.
4. We always backtracked to the last visited room.

5. There may be many possible solutions or there may be none.

6. There may be many shortest paths. In our example there are two: [1,3,4,6,7,8] and [1,3,4,5,7,8].

7. A maze may have cycles; in our case we have one cycle: [4,5,6,7].

Suppose a maze is stored in memory (and there are many ways to do so). What information do we have to keep on our journey?

1. We need to know if we have visited all the rooms. Otherwise if the solution does not exist, our search would never end. So we have to keep a boolean variable **finished**.

2. For every room we need to know if it was already **visited**.

3. For every room we need to know all its doors.

Using all this information, here is a first attempt at a program:

Illustration 12.1 *Travel through a maze.*

```
testNext[room_]:=
  Module[{finished = False},
        If[visited[room]], Return[finished],
          While[existsCorridor[room] && !finished,
            If[neighbor[room] != "B",
              testNext[neighbor[room]],
              finished = True; Return[finished]
            ]
          ]]]
```

In this program we introduce a new reserved word (spell): Return[<value>], or Return. This order ends the operation of the current function immediately and delivers **value**. We have never used Return before, because it destroys the natural order of behavior of the function, which should have one entry and one exit. Return is never needed, but might be useful for the readability and clarity of the program, because it allows us to check special cases and exit early.

Notice how recursion solves the problem of blind alleys and the problem of many doors. We will always use recursion in the solution of backtracking problems.

We will deal with two types of backtracking algorithms:

1. Finding all possible solutions.

2. Finding any one solution.

Finding all possible solutions

The search for all possible solutions has to be conducted in a systematic way, so that each solution is presented only once. One possible scheme for finding all solutions is very similar to the previous program, and might look as follows:

Illustration 12.2 *Search for all solutions by backtracking.*

```
next[i_Integer]:=
  Module[{j},
        For[j = 1, j <= m, j++,
            (* choose the j-th candidate *)
            If[permissible, (* mark it *)];
            If[i < n, next[i+1],
               (* print the solution *);
               delete[marked];
              ]
           ]]
```

In this program i denotes the next sequential try, n the number of possible tries, and m the number of candidates at each step.

Obviously we have to substitute appropriate code for the comments. We will use as an example a classical problem given in most texts describing backtracking, namely the "eight queens problem." To the best of my knowledge Gauss in 1850 was the first who dealt with this problem, but he did not find all the solutions. The problem may be formulated as follows: *How many ways can we place eight queens on an 8×8 chessboard in such a way that no two queens attack each other?* This means that no two queens may be on the same row, column, or diagonal.

Since a queen attacks every piece on the same column, there can be only one queen on a column. We choose a coordinate system that will make it possible to evaluate quickly the presence or absence of a queen on the two diagonals and in the same row.

Let i denote the index of the column, and let

1. $x[[i]]$ denote the position of a queen in the i-th column
2. $norow[[j]]$ denote the absence of a queen in the j-th row
3. $nolc[[k]]$ denote the absence of a queen on the k-th left diagonal ◢
4. $norc[[k]]$ denote the absence of a queen on the k-th right diagonal ◣

We also note that by a careful choice of the subscripts k of the diagonal tables $nolc[[k]]$ and $norc[[k]]$, all the places on the left diagonal ◢ have the same sum of the coordinates $i+j$, and all the places on the right diagonal ◣ have the same difference $i-j$.

The instruction to place a queen (in our scheme it is the j-th candidate) may be written as

$$x[[i]] = j;\ norow[[j]] = nolc[[i+j]] = norc[[i-j]] = False$$

and an instruction to delete a queen may be written as

$$norow[[j]] = nolc[[i+j]] = norc[[i-j]] = True$$

The final program might look as follows:

Illustration 12.3 *Placement of n queens on a chessboard.*

```
nn = 8;
x = Table[0,{nn}]
norow = Table[True,{nn}]    (* from 1 to 8 *);
nolc = Table[True,{2*nn}]   (* from -7 to 7 *);
norc = Table[True,{2*nn}]   (* from 2 to 16 *);

next[i_]:=
  (* Place 8 queens on a chessboard *)
  Module[
    {j},
```

```
    For[j = 1, j <= nn, j++,
        (* Choose the j-th candidate*)
        If[norow[[j]] && nolc[[i + j]]
            && norc[[i - j + nn]],
            x[[i]]=j;
            nolc[[j]]=norc[[i+j]]=norow[[i-j+nn]]=False;
            If[i < nn, next[i + 1],
                "Print the solution"];
            norow[[j]]=nolc[[i+j]]=norc[[i-j]]=True
]]]
```

I leave the completion of the program (printing the solution, or counting the number of solutions) as an exercise.

As an illustration of the program, I present a trace of the case of four queens on the 4 × 4 board.

position 1	position 2	position 3	position 4	position 5

position 6	position 7	position 8	position 9	position 10

In the above picture the letter X means a possible queen's placement. That is, a placement that does not cause immediate conflicts. The letter Y denotes an attempt at a placement with an immediate conflict.

As we can see by position 3, we cannot place a queen in the third row, so we have to backtrack. We backtrack again in position 5 (from the 4th row), and on the 6th position we conclude that placement of the queen in the upper left corner is impossible. The queen's placement in the second column leads to a solution shown in position 10.

There is a second solution that starts with the 3rd column that is symmetric in its vertical axes to the first solution.

One solution search

It would seem that looking for one solution is an easier task: you simply stop after you find one solution. Usually this is not so, because we do not assume the existence of a solution and we can finish our search prematurely. Therefore, the one-solution program is more complicated than the many-solution program. A general search scheme might look as follows:

Illustration 12.4 *One-solution search.*

```
next[i_Integer]:=
  Module[{j, suc = True},
         For[j = 1, suc && (j <= m), j++,
             (* choose j-th candidate *)
             If[acceptable, (* write it *)];
             If[i < n, next(i+1);
                If['not suc', 'delete it']
             ]
         ]]
```

As an example we will consider the problem of the knight's tour: given an $n \times n$ chess board and an initial knight's position, find a knight's tour such that each square on the board is visited exactly once.

Euler worked on this problem in 1758. We also know that for $n > 4$, there are always solutions.

This problem is interesting because a knight's movements do not have an obvious formula, therefore we will tabulate all moves a knight may make from a given point.

Let a knight's coordinates be x, y. From this position the knight can move in at most eight ways $x + dx[[i]]$, $y + dy[[i]]$, where $dx = \{2, 1, -1, -2, -2, -1, 1, 2\}$ and $dy = \{1, 2, 2, 1, -1, -2, -2, -1\}$.

I say at most since we cannot let the knight go outside of the board. The program might look as follows:

Illustration 12.5 *Knight's tour 1.*

```
n=5;
success = False;
xcoor = {2,1,-1,-2,-2,-1,1,2};
ycoor = {1,2,2,1,-1,-2,-2,-1};
board = Table[0,{i,n},{j,n}];

next[i_Integer, x_Integer, y_Integer]:=
  Module[
     {k = 0, xnext, ynext},
     While[!success && (k != 8),
           (* choose new candidate *)
           k++,
           xnext = x + xcoor[[k]];
           ynext = y + ycoor[[k]];
           If[xnext > 0 && ynext >0
              && xnext <= n && ynext <= n,
              (* move is on the board *)
              If[(* free place *)
                 board[[xnext, ynext]] == 0,
                 (* write the move *)
                 board[[xnext,ynext]] = i;
                 If[i < n*n,
                    next[i+1, xnext, ynext];
                    If[!success,
                       (* delete the move *)
                       board[[xnext, ynext]] = 0],
                    success = True
                 ];
              ] (* If free place *)
           ] (* If on the board *)
     ] (* While *)
  ] (* Module *);

board[[1,1]] = 1;
next[2,1,1];
Table[board[[i,j]],{i,n},{j,n}];
```

For $n = 5$, the execution time on my machine was 1.4 seconds, but for $n = 6$ the execution time was already 39.8 seconds. If we assume the time $T = A * x^{n*n}$ and use the time for $n = 5$ and $n = 6$, we estimate the time for $n = 7$ as $(1.36)^{49}$ seconds, many millions of years.

Let us try to speed this program up. As the first improvement let us change the matrix $board[[i, j]]$ into $board[[k]]$ where $k = (i-1)*n+j$, so we renumber this matrix by rows. Let xy be a number determining a knight's position. The knight in position xy may have 8 moves:

$$2 - n,\ 1 - 2n,\ -1 - 2n,\ -2 - n,\ -2 + n,\ -1 + 2n,\ 1 + 2n,\ 2 + n$$

which, when added to xy, give the knight's next position. We also add four columns and four rows of -1 to simplify checking 'on the board'. With these changes the previous program becomes:

Illustration 12.6 *Knight's tour 2.*

```
next2[i_, xy_]:=
  Module[{k = 0, move, p},
        While[!success && (k != n),
              (* choose k-th candidate *)
              k++;
              move = xy + addxy[k];
              p = board[[move]];
              If[p >= 0, (* move on the board *)
                If[p == 0, (* empty place *)
                  board[[move]] = i; (* write it *)
                  If[i < n * n,
                    next2[i+1, move];
                    If[!success, (* move failed *)
                      (* delete move *)
                      board[[move]] = 0],
                    success = True];
        ]]]]
```

I leave the finishing details for the reader. Here I show placing a 5×5

matrix, and the results of 5 × 5 and 6 × 6.

$$\begin{pmatrix} 1 & 2 & 3 & 4 & 5 \\ 6 & 7 & 8 & 9 & 10 \\ 11 & 12 & 13 & 14 & 15 \\ 16 & 17 & 18 & 19 & 20 \\ 21 & 22 & 23 & 24 & 25 \end{pmatrix}$$

Matrix numbered as a vector

$$\begin{pmatrix} 1 & 6 & 15 & 10 & 21 \\ 14 & 9 & 20 & 5 & 16 \\ 19 & 2 & 7 & 22 & 11 \\ 8 & 13 & 24 & 17 & 4 \\ 25 & 18 & 3 & 12 & 23 \end{pmatrix}$$

Results for a 5 × 5 matrix

$$\begin{pmatrix} 1 & 16 & 7 & 26 & 11 & 14 \\ 34 & 25 & 12 & 15 & 6 & 27 \\ 17 & 2 & 33 & 8 & 13 & 10 \\ 32 & 35 & 24 & 21 & 28 & 5 \\ 23 & 18 & 3 & 30 & 9 & 20 \\ 36 & 31 & 22 & 19 & 4 & 29 \end{pmatrix}$$

Results for a 6 × 6 matrix

The timing for the changed program on my machine was 1.1 seconds and 31.3 seconds for the 5×5 and 6×6 matrices, respectively. We have shortened the duration of the program about 25%, which in any program would be impressive, but here makes no practical difference for larger n. To understand our failure better, I added a counter of backtracks lb. For $n = 5$, lb is 8815, but for $n = 6$, lb is 248133, 28 times bigger. So we spend most of the time on backtracking.

There are many ways to solve the knight's tour problem, and you should look them up on the internet. But we are interested in the general problem of backtracking. Here we discuss a more general method of program shortening. We will do it in two steps:

1. To cut the number of attempts to find the free place on the board, we will construct a table of reachable places for every place on the board. This number is quite large, but smaller

then the 8 places we check. For example from the corner there are only two possible moves, the rest of them goes outside the board. An additional advantage is the possibility of skipping the test to determine if we are on the board. Then we do not have to increase the size of the matrix by four columns and four rows.

2. We notice that using the table of all possible moves cannot change the timing of the knight's tour substantially. It can shorten it, say, another 20%, but shortening the process from one week to 5 days will not bring us closer to the solution of the problem, so we have to find some other way. Looking carefully at the table, we realize that there must be ways to accomplish our tour without any backtracking. For a given n there must be at least as many tours without backtracking as there are solutions. The question is how to transform the given table in such a way as to minimize the amount of backtracking. For this purpose let us imagine a maze with many paths from point A to point B. To reduce the number of blind alleys we will choose paths that go through the smallest number of exits first. In the case of a knight's tour, we will choose the corners of the square (2 exits) first, then the edges, and only as last resort choose the middle of the square.

The next picture shows the original table of the knight's possible moves and the transformed table. The first column of both tables indicates how many moves are possible; further fields in the same row indicate where to go. The first row says that we have 2 moves: to the number 8 and to the number 12, which both have 6 exits, so we leave them alone. The second row has 3 moves: 11, 9 and 13, and was changed to 13, 11, 9 by sorting by first entries of the followers which are 4, 4, 8. Such static sorting was done for the whole table.

In the Footsteps of Programming Teachers

1	2	8	12					
2	3	11	9	13				
3	4	12	6	10	14			
4	3	13	7	15				
5	2	14	8					
6	3	3	13	17				
7	4	16	4	14	18			
8	6	17	11	1	5	15	19	
9	4	18	12	2	20			
10	3	19	13	3				
11	4	2	8	18	22			
12	6	21	1	3	9	19	23	
13	8	22	16	6	2	4	10	20	24
14	6	23	17	7	3	5	25	
15	4	24	18	8	4			
16	3	7	13	23				
17	4	6	8	14	24			
18	6	21	11	7	9	15	25	
19	4	22	12	8	10			
20	3	23	13	9				
21	2	12	18					
22	3	11	13	19				
23	4	16	12	14	20			
24	3	17	13	15				
25	2	18	14					

2	8	12						
3	13	11	9					
4	12	14	6	10				
3	13	7	15					
2	14	8						
3	13	3	17					
4	14	18	16	4				
6	17	11	15	19	1	5		
4	18	12	2	20				
3	13	19	3					
4	8	18	2	22				
6	3	9	19	23	21	1		
8	22	16	6	2	4	10	20	24
23	17	7	3	3	5	25		
4	18	8	24	4				
3	13	7	23					
4	8	14	6	24				
6	11	7	9	15	21	25		
4	12	8	22	10				
3	13	23	9					
2	12	18						
3	13	11	19					
4	12	14	16	20				
3	13	17	15					
2	18	14						

Knight's possible moves before and after transformation.

The method explained here is quite general for backtrack problems where there are many solutions and some paths have fewer exits than others. For the knight's tour, this method turns out to be very fast.

The table below gives the times and the number of backtracks for the knight's tour for $n \times n$ boards for $n = 5, 6, 7, 8, 9$ and 10.

n	5	6	7	8	9	10
time	0sec	0sec	.14sec	8.14sec	18.38sec	.016sec
bktr	0	11	1446	74544	156599	23

The time complexity seems to be completely unpredictable.

The final version of the knight's tour program is given below.

Illustration 12.7 *Faster knight's tour.*

```
next3[ii_, xy_]:=
  Module[
    {k, move},
    k = p[[xy, 1]] + 1;
    While[!success && (k > 1),
         (* Choose k-th candidate *)
         k--; move = p[[xy, k+1]];
         If[board[[move]] == 0, (* place free *)
            (* write it *)
            board[[move]] = ii;
            If[i < n * n,
               next3[ii+1, move];
               If[!success, (* failed try *)
                  (* delete move *)
                  board[[move]] = 0],
               success = True];
         ];]
  ];(* Module *)

board = Table[0, {i, n*n}];
board[[1]] = 1;
next3[2,1];
```

By default, *Mathematica* does not allow deeper recursion than 1024, so for this program to work for large n, we have to increase the system constant $RecursionLimit.

Chapter 13

In the footsteps of Keith Clark and Robert Kowalski,

inventors of Prolog, a language used in rule-based programming

(Rule-based programming)

We have discussed two programming paradigms: functional and imperative. Now we introduce a third: programming by writing general rules without being specific about which rule to use for the solution of a given case. The machine will use these rules to steer a computation. It has a choice of different rules to use to solve a given problem.

This paradigm raises the level of abstraction. Imperative programming using replacement, loops and choice constitutes the lowest level of abstraction, because we have to give orders to execute every minute

111

part of a computation. By using functions we raise the level of abstraction, since a function can describe much bigger chunks of a computation without worrying about how the computation is done. It is enough to know what to compute. Giving general rules about what to do in every case is the highest level of abstraction.

The simplest way to write rules is to use the same name for a whole group of functions. Using the same name for many different functions is called **polymorphism**. We used polymorphism already many times. For example, the function Plus[x,y] is different for different types of x and y; the machine chooses the one suitable for each data type. Polymorphism can be achieved in different ways: types of arguments, number of arguments, etc. We will concentrate on limiting the domain of arguments. We used domain limiting in the past as in f[n_Integer?Positive]. The question mark meant 'exclusively for' (positive integers). The ? operator limited the domain of the parameter n to positive integers. We could use the ? operator to define rules. For example, we could define absolute value as follows:

```
abs[x_?Negative]    := -x
abs[x_?NonNegative] := x
```

We define the function abs twice: once for Negative x and the second time for NonNegative x. Notice that the boolean functions Negative and NonNegative must already be defined (in this case by *Mathematica*'s library). An observant reader may ask what we gained by writing two functions instead of one:

```
abs[x_] := If[x<0, -x, x]
```

In the case of only two alternatives probably nothing, but in general we can have many alternatives. Writing many alternatives as nested If statements can be messy, unreadable and error prone. We can gather many rules into a set of rules, and leave it up to the machine to decide when to use a particular rule.

The use of the question mark operator is not very convenient because we have to define every function's limiting domain separately. It is more convenient to use a different operator: '/;' ('provided that'), which allows us to give the limiting condition immediately. This way we do not have to define any functions or give them any names.

As an example let us define differentiation of a polynomial of one variable named x. Below is a list of differentiation rules, with a worked example.

1. Derivative of sum.
$$d[f\_ + g\_] := d[f] + d[g] \qquad (13.1)$$
We check: $d[x^2 + 5x]$ gives $d[x^2] + d[5x]$.

2. Derivative of a product.
$$d[f\_ \ g\_] := f \ d[g] + g \ d[f] \qquad (13.2)$$
We check again: $d[x^2 + 5x]$ gives now $d[x^2] + 5d[x] + x \ d[5]$.

3. Derivative of a constant.

    Let a boolean function FreeQ[<expr>,<var>] give True for any expression <expr> independent of an object <var>. Using this function we write:
$$d[a\_] := 0 \quad /; \quad FreeQ[a, x] \qquad (13.3)$$

4. Derivative of the variable x.
$$d[x] := 1 \qquad (13.4)$$

5. Derivative of a power.
$$d[f\_ \wedge n\_] := n \ f \wedge (n-1) \ d[f] \quad /; \quad FreeQ[n, x] \qquad (13.5)$$
We check again: $d[x^2 + 5x]$ gives $5 + 2x$.

In five rules we taught a machine to differentiate a polynomial with respect to a variable x, letting the machine itself decide which functions (rules) to use.

The first two rules were for arbitrary functions f and g. The machine could decide which rule to use, since after d[] there were two different functions: a sum or a product. The third rule could be used when an arbitrary object a is not dependent on x. The fourth rule

$d[x] = 1$ was given only for a special variable named x. The fifth rule is similar to rules 1 and 2.

Before we congratulate ourselves, let us do some more checks of the function d for polynomial differentiation:

$$d[(5x+1)(x+1)] \quad \text{gives} \quad 1 + 5x + 5(1+x) \qquad (13.6)$$

which is true, but not simplified. Next:

$$d[2x-1] \text{ gives } 2 \qquad (13.7)$$

$$d[(5x+1)/(x+1)] \text{ gives } \frac{5}{1+x} - \frac{1+5x}{(1+x)^2} \qquad (13.8)$$

and

$$d[(5x+1)\text{Sqrt}(x+1)] \text{ gives } 5\sqrt{1+x} + \frac{1+5x}{2\sqrt{1+x}} \qquad (13.9)$$

Where did *Mathematica* get all this information about the derivative of a difference, a quotient, and a square root? Apparently we have unknowingly taught the machine more than we have written. In this case the results are correct, though unexpected, and unexpected results are always dangerous. These results appeared because *Mathematica* does preprocessing and uses certain internal representations which transform each division into a multiplication by a denominator taken to the power minus one, each subtraction into addition of a complemented term, and each square root into the power of a half. To find this out we will use a function `FullForm`.

    FullForm[a-b, a/b, Sqrt[x]]

prints

    [List[Plus[a, Times[-1, b]], Times[a,
        Power[b, -1]], Power[x, Rational[1, 2]]]]

which in our infix notation means

$$\{a + (-1)*b, \quad a*b^{-1}, \quad x^{1/2}\}$$

We did not need to know this representation until we defined rules to transform one expression into another. Now this lack of knowledge may be a source of trouble.

Notice that using a variable named x is dangerous because the same name might have been used before for other purposes. Also, it is not general because we cannot differentiate with respect to any other variable. To solve this problems we collect all of these rules into one function with a parameter x and obtain:

Illustration 13.1 *Derivative of a polynomial.*

```
der[pol_, x_]:=
  (* Compute derivative of polynomial with respect to x *)
  Module[{d},
         d[f_ + g_] := d[f] + d[g];
         d[f_   g_] := f d[g] + g d[f];
         d[a_] := 0   /; FreeQ[a,x];
         d[x] := 1;
         d[f_ ^ n_] := n f ^(n-1) d[f] /; FreeQ[n,x];
         d[pol]];

der[x^2 + y^2, x]   2x
```

To write more sophisticated programs with rules, we have to learn a little more about the use of multivariate functions: what can constitute a parameter and what a function? Parameters as well as results of functions can be variable names, arrays, lists, or other functions. Let us repeat what we have learned about functions up to now.

A multivariate function has the form `f[x1_,x2_,...,xn_]`. Here `f` is the name of the function defined for the arbitrary parameters x_i. Recall the programs for computing Fibonacci numbers. In the imperative paradigm, we used a computational loop. In the functional paradigm, we used recursion. In the rule paradigm, we write as follows: For boundary conditions:

$$Clear[f]; \quad f[1] = 1; \quad f[2] = 1; \qquad (13.10)$$

The function $Clear[f]$ deletes any machine knowledge of the object f. For other values of n we write as we did previously in functional style

$$f[n\_] := f[n-1] + f[n-2] \qquad (13.11)$$

If we want to compute this function in a 'lazy' way, keeping all the previously computed values, we change the last rule (13.11) to

$$f[n\_/;(n>2)] := f[n] = f[n-1] + f[n-2] \qquad (13.12)$$

Let us study the computation of $f[4]$ using rules (13.10) and (13.12). The machine checks the definition of the function f and finds only $f[1]$ and $f[2]$. Therefore it uses (13.12):

$$f[4] := f[4] = f[3] + 1, \quad f[3] := f[3] = f[2] + f[1] = 2 \qquad (13.13)$$

$f[3]$ is stored and used for the computation of $f[4]$, which is again stored. If we use the function f in the future, $f[i]$ will be used for $i = 1, 2, 3, 4$ and only for $i > 4$ will the function f be computed.

To describe more interesting uses of rules, we have to relax our concept of a function to allow a variable number of parameters.

In *Mathematica*, Blank[] (or _) is a pattern accepting a single arbitrary expression; $x\_$ accepts an arbitrary expression and defines its name as x. The pattern $x\_\_$ (two blanks) accepts one or more expressions and $x\_\_\_$ (three blanks) accepts zero or more expressions. For example given $f[x\_, y\_\_\_] := x + y$, then $f[a, b, c, d]$ gives $a + b + c + d$ and $f[a]$ gives a.

Let us study how *Mathematica* substitutes values of parameters in the case of many blanks. We write:

f[x___,y___] := Nil /; Print[{x},{y}];

f[1,2,3]
{},{1,2,3} {1}{2,3} {1,2}{3} {1,2,3}{} f[1,2,3]

This strange function gives Nil (the empty constant) and since the function Print is always *True*, it prints the values of the list x and the list y: all possible parameter substitutions.

A second generalization of function is to allow the definition of a function without a name, similar to Church's λ function. *Mathematica* has two ways to do this. One is $Function[< name >, < body >]$; the other where we give only <body>, using expressions where a symbol # stands for a parameter name. In the case of several parameters we use #1,#2, etc. We always end the definition of a nameless function with a symbol &.

These two definitions are equivalent:
$$distance[x\_, y\_] := Sqrt[x^2 + y^2]$$
$$distance = Sqrt[(\#1)^2 + (\#2)^2]\&$$

Here is an example of a function in which one of the parameters is a function and the result is also a function:

Illustration 13.2 *Derivative's approximation.*

```
derNum[f_, h_]  := Function[x, .5(f[x+h] - f[x-h]) / h];
```

```
derNum[#^2&, 0.001][1.]   2.0
```

Functions *Map* and *Apply* are especially important in functional programming, and are well suited to be used in nameless functions.

We have used *Map* before. In general $Map[f, g[x_1, x_2, \cdots, x_n]]$ produces $g[f[x_1], f[x_2], \cdots, f[x_n]]$. For instance Map[#^2-1&,{-1,2,3}] gives $\{0, 3, 8\}$. *Map* is used most often with the function $g = List$.

The function $Apply[f, g[< expr >]]$ produces $f[< expr >]$; it cuts out the name g. Again it is used most often with $g = List$. For instance $Apply[f, \{a, b\}]$ gives $f[a, b]$, $Apply[Plus, \{a, b, c\}]$ gives $a + b + c$, and $Apply[List, a + b + c]$ gives $\{a, b, c\}$.

Map and *Apply* are very useful in producing abstract rules, because they concentrate on the 'what' instead of on the small steps of 'how.'

We will now give a few examples of rule–based programs.

Illustration 13.3 *List's length.*

```
len[{ }] = 0;  len[li_List] := len[Rest[li]] + 1;
```

```
len[{1,2,5,0}]   4
```

This function redefines the function *Length*. It obtains the length of a list by adding one to the length of the tail of the list (*Rest*).

Illustration 13.4 *Sum of list's elements.*

```
sum1[li_List] := Apply[Plus, li];
sum1[1,2,3,4]   10
```

```
sum2[{ }]=0;  sum2[{first,re___}]:= first + sum2[{re}];
sum2[{1,2,3,4}]   10
```

*sum*1 is obvious. *sum*2 divides the list into the first element and the rest. It is a typical example of a rule–based program: division of a set into subsets and application of rules to each subset.

Illustration 13.5 *Bubble Sort.*

```
s[{x___,a_,b_,y___}]:= s[{x,b,a,y}] /; b > a;
sort[{ }] = { }; sort[{x_}]={x};
sort[li_List] := First[s[li]];

sort[{1,3,-2,4,0}]  {4,3,1,0,-2}
```

The function *sort* has a list as a parameter and produces a sorted list. We start by defining the auxiliary function $s[<list>]$ where $<list>$ has at least two elements a and b. Once called, s conditionally changes the order of elements a and b until the list is sorted. The function *sort* has 3 rules: for an empty list, for a list with one element, and for a list of at least two elements. The function $First$ cuts out the head s and gives the list itself.

Appendices

Appendix A

In the footsteps of John Backus and Peter Naur,
who invented a notation for describing the syntax of programming languages
(BNF syntax description)

The description of a programming language has to be clear and precise, because the machine always executes what we write and not what we have in mind. Every language contains at least two parts: a **syntax** or grammar of the language, which tells us if a given sentence belongs to the language, and a **semantics**, which describes a meaning of the sentence. In the case of a programming language, the semantics of a program is the action the machine has to take to properly execute a sentence. In this chapter we will deal with the syntax only. For instance a language describing a date might have the syntax *dd.dd.dddd*, where 'd' stands for an arbitrary digit. For instance, let us take the date 10.11.2009. The syntax of this sentence is correct, since it starts with two digits followed by a period, then two digits, a period, and four digits. However, its meaning or semantics is different in Europe (tenth of November 2009) and in the USA (eleventh of October 2009), which means that the syntax describes two different dates. Note also that many dates, although syntactically correct, make no sense. For instance, 32.75.0001.

A language describing another language is called a meta-language. One of the most popular meta-languages is BNF (Backus-Naur Form, or Backus Normal Form). It describes the syntax of so called **context-free languages** in which every sentence can be deduced from an initial symbol (I), an alphabet (A), auxiliary symbols (S), and a description of the grammatical rules (R).

Constants of BNF are all the symbols of the alphabet A and their arbitrary sequences, for instance '123' or 'v14'. Variables of BNF are constants enclosed in special parentheses, the symbols $\prec$ and $\succ$. For example: $\prec variable \succ$. There are two important meta-symbols: '::=' – meaning 'is defined as' and '|' – meaning 'or'. The symbol '|' is used as an abbreviation for the enumeration of the elements of the same construct. For instance

$\prec sign \succ ::= +$ and $\prec sign \succ ::= -$

may be written as

$\prec sign \succ ::= + | -$

We define digits as

$\prec digit \succ ::= 0|1|2|3|4|5|6|7|8|9$

To describe an unsigned integer as a sequence of digits we use a recursive definition:

$\prec natural \succ ::= \prec digit \succ | \prec natural \succ \prec digit \succ$

We read this as 'natural may be a digit or a natural (already defined) followed by a digit.' This type of recursion is called left recursion, because the defined object is of the form

$\prec object \succ ::= \prec something \succ | \prec object \succ \prec anything \succ$

The defined object is on the left, and additions are concatenated to the right. Operations obtained by left recursion are often called **early**, and operations obtained by right recursion are called **late**.

BNF is a very powerful language because a few simple rules can define an infinite number of objects as in the case of the definition of a natural.

We can define an integer constant (int) the following way:

$\prec sign \succ ::= + \mid -$

and

$\prec int \succ ::= \prec natural \succ \mid \prec sign \succ \prec natural \succ$

For the clarity of some definitions it is useful to define the symbol $\prec nil \succ$, or its equivalent form ϵ, meaning the absence of any symbol. Using ϵ, we can define an integer constant as:

$\prec sign \succ ::= + \mid - \mid \epsilon$

$\prec int \succ ::= \prec sign \succ \prec natural \succ$

To describe the grammar of the sentence we use only constructs

$\prec name \succ ::= <\text{expression}>$

where expression uses two binary operations: '|' (alternative) and ' ' (followed by). Notice that since the operation 'followed by' has higher priority than '|' the definition

$\prec int \succ ::= + \mid - \mid \epsilon \prec natural \succ$

is incorrect. To make it right we need 'meta-parentheses', which BNF does not have. This omission causes the usage of many names, and makes grammatical definitions much longer.

Now we are ready to describe the syntax of a language called "Little Math".

1. ≺program≻   ::= ≺fundefs≻≺funuses≻
2. ≺fundefs≻   ::= ≺fundef≻ | ≺fundefs≻ ≺fundef≻
3. ≺funuses≻   ::= ≺funuse≻ | ≺funuses≻ ; ≺funuse≻
4. ≺fundef≻   ::= ≺name≻[≺listpar≻] :=≺instr≻
5. ≺funuse≻   ::= ≺name≻[≺listarg≻]
6. ≺listpar≻   ::= ϵ | ≺pars≻
7. ≺pars≻   ::= ≺name≻_ | ≺pars≻,≺name≻_
8. ≺listarg≻   ::= ϵ | ≺args≻
9. ≺args≻   ::= ≺expr≻ | ≺args≻, ≺expr≻
10. ≺expr≻   ::= ≺expr≻ ≺plusop≻ ≺term≻ | ≺term≻
11. ≺plusop≻   ::= + | −
12. ≺term≻   ::= ≺term≻≺mulop≻ ≺primary≻
13. ≺mulop≻   ::= * | | / | ϵ
14. ≺primary≻   ::= (≺expr≻) | ≺name≻ | ≺number≻
                        | ≺funuse≻ | ≺known≻
15. ≺instr≻   ::= **If**[≺boolean≻,≺expr≻,≺expr≻]
                        | ≺expr≻
16. ≺boolean≻   ::= ≺expr≻ ≺relop≻ ≺expr≻
17. ≺relop≻   ::= == | != | < | <= | > | >=
18. ≺known≻   ::= **Round**[≺*expr*≻]
                        | **Quotient**[≺*expr*≻,≺*expr*≻]

I leave the definitions of ≺var≻ and ≺number≻ to the reader.

Possible additions to the syntax include rules for an arbitrary number of spaces between sentences, treatment of the end of line as a space, and the addition of comments anywhere with the construct '(* <comment> *)'.

Context-dependent constructs, such as the fact that functions used have to be defined and that each parameter in a function definition has to have a different name, cannot be described using BNF.

We described arithmetic operations using left-recursion, because they are early. For instance $a + b + c = (a + b) + c$. If we added an operation for raising a number to a power we would use right-recursion, since mathematicians want $a \wedge b \wedge c$ to mean $a \wedge (b \wedge c)$.

The priority of the arithmetic operations is correctly described by rules 10 and 12.

Appendix B

In the footsteps of Gottfried Leibniz,
who in the 17th century taught us the binary representation of numbers

(Number representation)

The beginning of counting can be traced to the need of humans to know about their lives: the number of sheep or other cattle, the area of a farm's size, the passing of time. This need existed long before we learned to read and write. The only knowledge we have about these prehistoric times comes from paleontology and from pieces of old stones and bones. Mathematicians of those times used a wolf's bones rather than chalk and blackboard. A key discovery about European prehistoric mathematicians was Karl Absolom's finding in 1937 of a 30,000 year old wolf's bone on which were 55 cuts. This bone was the laptop of the stone age, used for counting and remembering numbers.

We don't know what the owner of this bone was counting (maybe cattle), neither do we know the name of the wolf, but this way of writing a number was very interesting. It let the owner check if the count was always the same without the need of naming the number. The method of cuts was improved by grouping the cuts, five in a group. Such a method of counting we call 'unary', because it uses only one symbol to denote any number. The operations of addition and subtraction are very simple: we concatenate 2 numbers for addition, and erase ones for subtraction. The only minus of such notation is the length of a number, but in those times big numbers were not used. In later times a more modern shepherd used different names for different groups: say f for a group of 5 and t for a group of 25, so the notation for 55 might look like ttf, and for 57 sheep– ttf11. In different cultures the group's sizes were different: 5, 10, 20, or even 60, but in later times the standard was 10.

A similar system is the Roman system, where the letters I, V, X, L, C, and M mean 1, 5, 10, 50, 100, and 1000 respectively. The only departure from the system just described is adding the prefix of I (one) to the left of a number with the meaning of a subtraction. For instance IV means four. We can write the consecutive numbers as I, II, III, IV, V, VI, VII, VIII, IX, X, and the year 2009 as $MMIX$.

The system we use now is of Arab–Indian origin. It is a positional system, because a change of a digit's position in the number changes its value. For instance 15 and 51 denote different numbers. In a positional system every natural number N has the form

$$N = d_k d_{k-1} \ldots d_1 d_0 = d_k b^k + d_{k-1} b^{k-1} + \cdots + d_1 b + d_0 \qquad (B.1)$$

Here the number b is called the base, and d_i are called digits, and satisfy the relation $0 \leq d_i < b$. The most common bases are $b = 10$, 2, and 16. The base $b = 10_D$ is called 'decimal' and is presently used everywhere. The binary base ($b = 2_2$) is used by computers, and binary digits are called bits.

The hexadecimal base ($b = 16_H$, also called 'hexa' or 'hex') is used by people using binary numbers who don't want to write very long sequences of bits (zeros and ones). For instance the number

$$100_D = 64 + 32 + 4 = 2^6 + 2^5 + 2^2 = 1100100_2$$

uses 7 digits, but grouping them 4 at a time from the end gives a number with only 2 digits: $64_H = 6*16 + 4$. In the hexadecimal notation we use decimal numerals up to 9 and then the consecutive letters a, b, c, d, e, f, as in the C language. We call a change of a number from one base to another a 'conversion'. The table below gives the conversion from binary to hexa:

$b=2$	0000	0001	0010	0011	0100	0101	0110	0111
$b=16$	0	1	2	3	4	5	6	7

$b=2$	1000	1001	1010	1011	1100	1101	1110	1111
$b=16$	8	9	a	b	c	d	e	f

The conversion table from binary to hexa.

To convert a number from hexa (or binary) to decimal we use the formula (B.1) and evaluate a corresponding polynomial. A conversion from decimal to binary can be done in two ways. One way is the top digits first, as we did previously, e.g.,

$$228_D = 128 + 64 + 32 + 4 = 2^7 + 2^6 + 2^5 + 2^2 = 11100100_2 = e4_H$$

subtracting the highest powers of two. The other way is to get the lower digits first, dividing by the base and gathering the remainders as the consecutive digits, for instance $228/16 = 14$ with the remainder of 4, so the last hexa digit is 4, and the one before the last is $14_D = e_H$.

In the binary representation negative integers are usually represented using two's complement. To obtain two's complement we perform a flip-flop, changing all the zeros in the number into ones and all ones into zeros, and then add one. For instance to obtain -1 in a 4-bit cell we take 0001_2, change it into 1110_2, and add 1 obtaining 1111_2.

Using such a representation we can add two integers the same way as positive integers, knowing only the addition table:

$$0+0=0, \quad 0+1=1, \quad 1+0=1, \quad \text{and} \quad 1+1=10$$

In comparison with the rules of addition learned at school, two's complement addition is much simpler. By the schoolroom method, for instance, to add $(-6) + (+5) = (-1)$ we have to see first if the signs are the same; if different, we have to determine which magnitude is bigger, subtract the numbers, and attach a sign from the number with the bigger magnitude. Compare this to binary addition in our 4-bit register:

$$1010 + 0101 = 1111$$

Two's complement representation introduces a nonzero number that behaves as zero, i.e., $-a = a$. In our 4-bit register it is 1000_2.

Some computers use one's complement representation in which only flip-flop is used, but then we have 2 zeros: all zero or all one bits.

We represent real numbers as floating point numbers in a computer. The floating point number $\mathcal{R}$ is represented as a triple: a sign S, an exponent E, and a mantissa M in such a way that $\mathcal{R} = M \cdot b^E$, where b is the base 2 or 16. After many different representations were tried the IEEE (Institute of Electrical and Electronic Engineers) introduced a standard in which we have three representations: a float of 32 bits, a double of 64 bits, and a quad of 128 bits.

The format of the float number is 1 bit for a sign, 8 bits for a weighted exponent (i.e., exponent + 01111111_2), and 23 bits for a normalized mantissa M, in which the first bit is omitted and $1 \leq M < 2$. The first bit of a mantissa is 0 for the number zero and 1 for all other numbers. The floating point number 0.0 is represented as an integer with all zero bits. For instance, the number 1.0_D will be

$$1.0_D = 0 + 01111111_2 + 00\ldots0_2 = 3f800000_H$$

The number 2.0_D will have an exponent greater by one, so it will be

$$2.0_D = 0 + 10000000 + 00\ldots0_2 = 40000000_H$$

Numbers -1.0_D and -2.0_D have different sign bits, so they will be $-1.0_D = bf800000_H$ and $-2.0_D = c0000000_H$.

To convert a fraction f into a float we have to find a binary representation of f. Again we can use multiplication or division. For instance to convert $f = 0.1_D$, we multiply f by 2^n, choosing n to make $1 < f \cdot 2^n \leq 2$. In this case $n = 4$. This gives us an exponent $= -4_D$. To obtain the mantissa we notice that if we multiply a number

$$f = b_{-1}2^{-1} + \cdots + b_{-k}2^{-k} + \cdots$$

by b, then the integer part of $f \cdot b$ gives us b_{-1}. Repeating this process, we can get as many bits of mantissa as we need. We obtain

$$1.6 = 1 + .6, \quad .6 * 2 = 1.2 = 1 + .2,$$

$$.2 * 2 = .4 = 0 + .4, \quad .4 * 2 = .8 = 0 + .8,$$

$$.8 * 2 = 1.6$$

Therefore the consecutive bits are 1.10011001100... which are repeated, so our mantissa after neglecting the leading 1 is $10011001100..._2$, and the whole number is

$$0 + 01111011 + 10011001100..._2 = 3dcccccd_H$$

where the last d was obtained by rounding.

The format for a double is 1 bit for the sign, 11 bits for the weighted exponent, and 53 bits for the mantissa. Since the number of bits for the sign plus exponent is 12, or 3 digits hex, the conversion from decimal to hex is simpler. Here are examples:

$$1.0_D = 0 + 011\ 1111\ 1111_2 + 00...0_2 = 3ff0\ 0000\ 0000\ 0000_H$$

$$0.1_D = 0 + 011\ 1111\ 1011_2 + 10011001...1010_2 = 3fb9\ 9999\ 9999\ 999a_H$$

$$-(1/3)_D = 1 + 011\ 1111\ 1101_2 + 01010101...0101_2 = bfb5\ 5555\ 5555\ 5555_H$$

The numbers of type half and quad are very seldom used. They have the form

$$\text{half}(1 + 5 + 10) \quad \text{and} \quad \text{quad}(1 + 15 + 112)$$

In addition to floating point numbers there exists bit combinations giving the 'numbers' $+\infty$, $-\infty$, and 'undetermined'. We will not deal with them here.

Since floating point numbers are only an approximation to real numbers, we commit an error due to the truncation. The accuracy of these different representations can be estimated using $a[n\_] := 2^{-n}$. This gives $a[11] = 5.10^{-4}$ for a half, $a[24] = 6.10^{-8}$ for a float, $a[53] = 1.10^{-16}$ for a double, and $a[113] = 1.10^{-34}$ for a quad.

Appendix C

In the footsteps of Steven Wolfram,
inventor of Mathematica, the language used in this book

(Introduction to Mathematica)

I use *Mathematica*, the language invented by Steven Wolfram, in all my example programs. It is one of the richest languages used by mathematicians and engineers.

At the beginning of the 20th century, every engineer used a slide rule for computing. In the middle of the 20th century, mathematical tables were in use; later a calculator was ever-present. In the 21st century, it is impossible to imagine an engineer without a laptop and some algebra system such as *Mathematica*, *Maple*, or *Matlab*.

The main difference between an imperative language and an algebra system is the treatment of variables. In C–like languages the expression $x = a + b$ makes sense only if the variables a, b, and x are defined. In algebra systems the value of the variable x is the algebraic sum $a + b$. If any variable is not initialized its value is the name itself.

In the Footsteps of Programming Teachers

The capabilities of *Mathematica* include: arbitrary precision arithmetic, operations on algebraic expressions such as expansion and factoring, calculus, operations on power series, 2 and 3 dimensional color graphing, matrix and tensor algebra, solving of algebraic and differential equations, and numeric calculations, among others.

Constants may be arbitrarily large integers, rationals of the form $int1/int2$, real (floating point) with arbitrary precision, complex of the form $a + I\,b$, and symbolic constants such as $Log[2]$, Pi, or $Sqrt[5]$.

In *Mathematica*, function application uses brackets '[ ]' rather than parentheses '( )', which are used only for grouping. Braces '{ }' are used to define a list. Values are assigned using '=' and function definitions using ':='. The names of built-in functions begin with a capital letter.

Mathematica starts its operation with 'In[1]:=' and waits for some computation. The result of this computation may be used later as the value of the variable %1. As the computation progresses, *Mathematica* increments n in $In[n]$. If the expression to evaluate is not terminated with the ';' symbol, *Mathematica* prints the result of evaluation.

As an example, suppose we have a function $f(n) = n^2 + \sqrt{n}$.
To define this function in *Mathematica* we write

```
f[n_]:= n ^ 2 + Sqrt[n]
```

Here the underline sign after n means "for arbitrary" n, the sign ':=' means "is defined", and *Mathematica* promises execution as soon as we call this function, as in $f[0]$.

The variable n may even be a complex number, as in $f[1 + 3I]$. If we want to compute f for n integer only, we write

```
f[n_Integer]:= n ^ 2 + Sqrt[n]
```

Then if by mistake we write $f[1.5]$ this function definition is not used, and f is not computed.

The same function may be defined as many times as we wish, and all of these definitions will exist at the same time. If we want to start definitions from scratch we write `Clear[f]`.

We may further limit Integer n to the nonnegative numbers by writing

```
f[n_Integer?Nonnegative]:= n ^ 2 + Sqrt[n]
```

We may compute the same function using an expression

```
expr = n ^ 2 + Sqrt[n]
```

This expression is not a function, so $expr[2]$ is not correct, but we can use a substitution in this expression of the value of n with something else for instance

```
t = expr /. n -> 2
```

We substitute ('/.') in expr the number 2 in place of n, and place the result in t. Notice that the value of n is unchanged in expr, and 2 is used for the computation of t, giving as a result $t = 4 + \sqrt{2}$.

If we want to see the approximate numeric result, we use the built-in function $N[t]$, usually yielding 6 decimal figures. The function SetPrecision[t, 10] always prints 10 digits.

Below we list the functions used in this book with short explanations and examples. We use the following notation: e or e_n denotes an arbitrary expression, be a boolean expression, ie integer expression, and it for an iterator repeating a given action. An iterator is a list and may have many forms. Some of them are:

1. a natural number denoting the number of repetitions. For example, $x = Table[0, \{5\}]$ defines x as a list of 5 elements (iterator) and sets each element to 0.

2. $i, imax$: i goes from 1 to $imax$.

3. $i, imin, imax$: i goes from 1 $imin$ to $imax$.

4. $i, imin, imax, d$: i goes from $imin$ to $imax$ in steps of d.

List of functions used

1. Apply, page 117.

    Apply[f1, f2[$\cdots$]] gives $f1[\cdots]$. It is usually used with $f2$=List and then changes $\{p1, p2, \cdots, pn\}$ into $f1[p1, p2, \cdots, pn]$.

    Example: Apply[Plus,1,2,3,4] gives 10.

2. Begin, page 44.

   Begin["name'"] changes the context of all objects defined from this moment until the function End[ ].

   Example: Begin["fibo'"] starts the new context fibo'.

3. Ceiling, page 49.

   Ceiling[x] gives the smallest integer not smaller than x.

   Example: Ceiling[Pi] -> 4.

4. End, page 44.

   End[ ] returns to the previously defined context. Every End[ ] must match a previously defined Begin.

   Example: End[ ] gives an error message and present context Global'.

5. DeleteCases, page 89.

   DeleteCases[list, n] deletes all occurrences of n from the list.

   Example: DeleteCases[1,2,1,1] gives {2}.

6. First, page 118.

   First[list] gives first element of *list*, and is the same as list[[1]].

   Example: First[5,2,3] -> 5.

7. Floor, page 49.

   Floor[x] gives smallest integer not larger than x.

   Example: Floor[Pi] -> 3.

8. For, page 16.

   For[*start, be, change, action*] executes *start*, and then repeatedly evaluates the triple (*be, action, change*) as long as *be* is *True*. The result of the function For is always *Null*.

   Example: For[sum=0; i=0, i<3, i++, sum += f[i]] gives the result $sum\ f[0] + f[1] + f[2]$.

9. FromCharacterCode, page 96.

    FromCharacterCode[list] changes every number n in the list ($0 \leq n \leq 127$) to text using ASCII representation.

    Example: FromCharacterCode[72,101,114,101] -> Here.

10. If, page 3.

    If[be, e_t, e_f] gives e_t if be is *True*; otherwise it gives e_f. If[be, e_t] gives *Null* for be *False*.

    Example: min = If[a < b, a, b].

11. Length, page 89.

    Length[list] gives the number of elements in the list.

    Example: Length[a, b, c] -> 3.

12. Map, page 96.

    Map[f, g[...]] exchanges the names f and g.

    Example: Map[f,a,b,c] -> {f[a],f[b],f[c]}.

13. Mod, page 84.

    Mod[p, q] gives the remainder on division of p by q. The result has the same sign as q.

    Example: Mod[5,2] -> 1.

14. Module, page 16.

    Module[$\{x, y, \ldots, z\}$, e] makes names $x, y, \ldots, z$ local to e. These variables may be initialized.

    Example:

    ```
    ss[f_]:= Module[{sum=0, i},
                For[i=0, i<3, i++, sum += f[i]];
                sum]
    ```

    gives $f[0] + f[1] + f[2]$ for arbitrary f, and the variables i and *sum* do not change the values of i and *sum* outside the function *ss*.

15. N, page 44.

    N[e] gives 6 significant digits of e. N[e, n] sometimes gives n digits of e.

    Examples: N[Pi] -> 3.14159, and N[Pi,10] -> 3.141592654.

16. OddQ, page 28.

    OddQ[e] gives True if e evaluates to an odd integer.

    Example: OddQ[3.0] gives False, OddQ[6/2] gives True.

17. Plus, page 3.

    Plus[a, b, ...] computes the sum $a + b + \ldots$

    Example: Plus[1,2,3] -> 6.

18. Power, page 114.

    Power[a,b] computes $a \wedge b$ the same way as a ^ b does.

    Example: Power[2,5] -> 32.

19. PowerMod, page 96.

    PowerMod[m, d, n] computes $m \wedge d$ mod n.

    Example: PowerMod[2,5,10] -> 2.

20. Quotient, page 4.

    Quotient[e1, e2] gives the integer part of $e1/e2$, defined as Floor[e1/e2].

    Example: Quotient[3.,2.] -> 1.

21. Rest, page 117.

    Rest[*list*] gives the *list* with the first element removed.

    Example: Rest[a,b,c] -> {b,c}.

22. Return, page 99.

    Return[e] ends computing of the current function prematurely and gives e. Return[ ] ends the computation and gives Null.

    Example: If[x<0, Return[1]].

23. Round, page 4.

    Round[e] gives the integer closest to e. If e is an integer+0.5, Round gives the closest even integer.

    Example: Round[2.5] -> 2 and Round[Pi] -> 3.

24. Sqrt, page 114.

    Sqrt[e] gives the square root of e.

    Example: Sqrt[2.] -> 1.41421 and Sqrt[-1] -> I.

25. Table, page 42.

    Table[e, {it}] generates a list using the iterator *it*. It builds a list, matrix, or tensor.

    Examples:

    ```
    Table[1, {3}] -> {1,1,1}
    Table[f[i], {i,3}] -> {f[1],f[2],f[3]}
    Table[i-j, {i,2}, {j,2}] -> {{0,-1},{1,0}}}
    ```

26. Times, page 114.

    Times[$a, b, \ldots$] computes the product $a * b * \ldots$

    Example: Times[1,2,3] -> 6.

27. While, page 48.

    While[*be*,*action*] is the same as For[, *be*, , *action*]. It repeats the *action* as long as *be* is True.

    Example: sum=0; i=0; While[i<3, sum += f[i]; i++] gives in the variable *sum* $f[0] + f[1] + f[2]$.

Appendix D

In the footsteps of the Babylonians, *who around 1800 BC showed us how to solve quadratic equations*

(**Quadratic equations**)

The oldest quadratic equation solution method was found on an ancient Babylonian clay tablet. This tablet describes how to solve the system of equations:

$$x + y = p, \qquad xy = q \qquad \text{(D.1)}$$

which is equivalent to the quadratic equation

$$x^2 + q = px \qquad \text{(D.2)}$$

In the Footsteps of Programming Teachers

The Babylonian tablet contained the following algorithm to solve the system (D.1):

1. Compute $\dfrac{x+y}{2}$

2. Compute $\left(\dfrac{x+y}{2}\right)^2$

3. Compute $\left(\dfrac{x+y}{2}\right)^2 - x\,y$ \hfill (D.3)

4. Compute $\sqrt{\left(\dfrac{x+y}{2}\right)^2 - x\,y} = \dfrac{|x-y|}{2}$

5. Compute x, y using 1 and 4.

Example:
Solve the equation $x^2 + 3 = 4x$, where $p = 4$, $q = 3$.

After the first step we obtain 2; after the second step, 4; after the third step, 1; and after the fourth step, 1. Therefore

$$\frac{x+y}{2} = 2, \qquad \frac{x-y}{2} = 1 \tag{D.4}$$

which gives $[x, y] = [3, 1]$.

The algorithm (D.3) can be traced to the same equation that I mentioned in Chapter 2, namely (2.8):

$$a \times b = \left(\frac{a+b}{2}\right)^2 - \left(\frac{a-b}{2}\right)^2 \tag{D.5}$$

This equation has a pretty geometric interpretation:

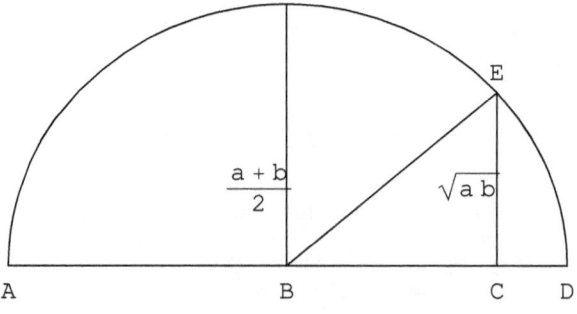

APPENDIX D. BABYLONIANS

On this picture the segment AC = a and CD = b. The circle's diameter is $a+b$, therefore BD=BE=$\frac{a+b}{2}$ and the segment BC=$\frac{a+b}{2} - b = \frac{a-b}{2}$. Equation (D.5) comes from the Pythagorean theorem for the BCE triangle.

Let us study the general solution of a quadratic equation

$$ax^2 + bx + c = 0 \tag{D.6}$$

for arbitrary real numbers a, b, c.

In spite of the fact that a program for solving this equation is given in most elementary books on programming, it is one of the harder small programming projects.

Let us write the standard solution of this equation (D.6):

$$x_{1,2} = \frac{-b \pm \sqrt{b^2 - 4ac}}{2a} \tag{D.7}$$

Keeping in mind the representation of real numbers $\mathcal{R}$ and the fact that each number r represents the interval $|r - r^*| \leq \epsilon$, let us ask what bad things can happen in our computations?

1. $a = b = 0$: there are no solutions, or every number is the solution.

2. $a = 0$, $b \neq 0$: one or no solutions.

3. $a \neq 0$, $D = b^2 - 4ac$: b^2 or $4ac$ may overflow.

4. $D < 0$: complex roots.

5. $D = 0$: 1 root.

6. $D > 0$, $b^2 >> 4ac$: one of the roots is close to zero, loss of accuracy.

7. $D > 0$, $b^2 \approx 4ac$: two roots very close to each other.

8. $b^2 << 4ac$: b^2 has no influence on the final result, even when b itself changes the accuracy

To resolve some of these difficulties let us recollect Vieta's formulae for the quadratic equation.

If x_1 and x_2 are the roots of the quadratic equation (D.6), then the following conditions are true:

$$x_1 + x_2 = -b/a, \quad x_1 x_2 = c/a \qquad (D.8)$$

Let us assume that there exists a real number $\mathcal{R} = \infty$, which we obtain when the result of any operation satisfies the condition

$$|F| > M \approx 10^{38} \qquad (D.9)$$

and that the arithmetic unit calls the operating system only when such a number ($\mathcal{R} = \infty$) is used. Let us assume further that there exists a boolean function `overflow[x]`, which gives True in case of overflow. Our aim is to solve one of these equations:

A. $ax^2 + bx + c = 0$, or

B. $cy^2 + by + a = 0$.

Since the second equation has the solution $y = 1/x$, solving either of these will solve our problem. We want to solve a simpler equation

$$x^2 + 2px + q = 0$$

which we get from dividing equation A by a or equation B by c, because it is easier to analyze the behavior of two constants p and q rather than three a, b, c. The plan of the solution might look as follows:

1. $a = 0$: no solutions, one solution, or every x is a solution.

2. overflow$[q = c/a]$: we solve equation B.

3. q is okay: check overflow$[p = .5b/a]$, one solution is large, the other small.

4. p and q okay: $|p| > \sqrt{M} = 10^{19}$ (see (D.9)), p^2 overflows, we compute $D1 = p\sqrt{1. - q/p/p}$.

5. p^2, q okay: we compute $D = p^2 - q$.

6. $D < 0$: complex solution.

7. $D \geq 0$: we compute $D1 = \sqrt{D}$.

8. We compute the root which has the sign of p, or $x_1 = -p + D1$ for $p < 0$, or $x_1 = -p - D1$, to avoid a bad subtraction.

9. We compute the second root from $x_2 = q/x1$.

This plan is not bad, but has some faults:

1. The plan is not complete: it does not cover all cases.

2. In case 8 we still may have a bad subtraction when both roots are close together, as in the equation $x^2 - 2x + 1 - \epsilon^2 = 0$, which has solutions $x_{1,2} = 1 \pm \epsilon$.

3. There are intervals which are not accurate, i.e., there exist solutions in $\mathcal{R}$ that we will not obtain. For instance the solution x may exist, but $1/x$ does not.

4. Reduction to the canonical form might give wrong results. For instance, let us take the equation

$$\eta x^2 - 3x + 2/\eta = 0, \quad \eta > 0$$

It has two solutions: $x_1 = 1/\eta$ and $x_2 = 2/\eta$. If η is small (say $\eta = 10^{-30}$), the machine treats η^2 as zero; therefore we will get one solution $x = 2/(3\eta)$. If η is large, say, $\eta = 10^{30}$, we will obtain $x = 3/\eta$. Not only did we lose one solution, but the other one is wrong.

As we see, in spite of our hard work our results are not satisfactory. We still have some options for correcting the plan:

1. Use the fact that the type double has a much bigger range than single precision since it uses 4 more bits for its exponent and do parts of the computation in double. This will solve our immediate problem, but does not say anything about how to solve the problem in the case a, b, c are double.

2. Substitute x_1 and x_2 into equation (D.6), and if we are far from zero, use numerical solutions. For example we can use one iteration of

$$x_{n+1} = \frac{ax_n^2 - c}{b + 2ax_n} \qquad \text{(D.10)}$$

This formula is a generalization of Heron's method for square root extraction.

3. Change the scale of numbers, so they will not be very small or very large. Consider a substitution $x = \eta y$. Then our equation (D.6) becomes
$$\eta^2 a y^2 + b\eta y + c = 0 \qquad (D.11)$$
If $a \neq 0$ and $c \neq 0$, we can choose
$$\eta = \sqrt{|c/a|} \qquad (D.12)$$
and obtain equation
$$y^2 + 2py + q = 0 \qquad (D.13)$$
where $q = \pm 1$ and $p = b/(2c)\eta = b/(2\sqrt{|ac|})$. We notice that if $a \neq 0$ and $c \neq 0$, then η always exists, even if c/a does not exist, and we can compute
$$\eta = 10^s \sqrt{|c \cdot 10^{-2s}/a|}$$
choosing two s values, one such that $|c/a|$ is not too large (giving overflow), and the second such that this quantity is not too small (causing a sudden zero).

We can simplify this analysis if we assume that for every x in $\mathcal{R}$ we have a function `exponent[x]` giving the exponent of x. We have
$$-126 \leq \text{exponent}[x] \leq 128$$
Let $\alpha = \text{exponent}[a]$ and $\gamma = \text{exponent}[c]$. From (D.12) we have
$$e = \text{exponent}[\eta] = (\alpha - \gamma)/2$$
which always exists, and we can use $\eta = 2.0^e$.

I hope this discussion convinces you that the solution of quadratic equations using a computer is not as simple as some people think.

Appendix E

Online judges

An online (programming) judge is a website where you are given a programming problem, you submit code to solve the problem, and that code is automatically judged, first for correctness, but perhaps also for speed, or code size, or error handling. An online judge will often support submissions in many different computer languages.

There are many online judges, offered by academic institutions, companies, and independent organizations. It is hard to give a list, because they vary in user experience, and the landscape is shifting. However, in 2021, commercial (but free to use) offerings include Coderbyte, HackerRank, or LeetCode; one academic offering is the UVa online judge; and one independent organization is Project Euler. For Project Euler, you don't submit code, just the answer. Every problem is written so the answer is a single value. If none of these are suitable, search the web or Wikipedia to find more.

People use online judges for many reasons: to learn, to have fun programming, or to show publicly that they are a good programmer. Getting good scores on some commercial sites can lead to job offers.

To try a judge, simply find the website and look for the problems, sometimes called "challenges". You might be able to try them even without creating an account. Give it a go!

Appendix F

Mathics

Mathics (at mathics.org) is a free, open-source alternative to Mathematica started in 2016. There are many ways to try it, described in their documentation.

One fairly easy way is to use `pip` (a package installer for the Python programming language) to install `mathicsscript`. Then you can run `mathicsscript` and use the interactive environment to run code. A short example is shown below,

```
$ pip install mathicsscript
... installation omitted

$ mathicsscript
Mathicscript: 2.1.2, Mathics 2.1.0
on CPython 3.9.1 (default, Dec 24 2020, 15:57:37)
using SymPy 1.7.1, mpmath 1.2.1, numpy 1.20.2

Copyright (C) 2011-2021 The Mathics Team.
This program comes with ABSOLUTELY NO WARRANTY.
This is free software, and you are welcome to redistribute it
under certain conditions.
See the documentation for the full license.

Quit by evaluating Quit[] or by pressing CONTROL-D.
```

In the Footsteps of Programming Teachers

```
In[1]:= abs[x_] := If[x < 0, -x, x]
Out[1]= None

In[2]:= abs[3]
Out[2]= 3

In[3]:= abs[-3]
Out[3]= 3
```

ACKNOWLEDGEMENTS

Many thanks to my son Daniel for spending many days correcting my mistakes, pushing me to help him, pushing my colleagues to read my book, and getting my niece Anetta Dobrakowska to paint the cover. He also wrote the last two appendices and checked many Mathematica programs in Mathics.

I also want to thank my wife Elaine who polished my translation from the Polish language. The book is a translation of the Polish version published in 2012 in Łódź.

Many thanks to my colleagues Stanisław Goldstein and Daniel Boley, who read the book and wrote valuable remarks. Thanks also to Anetta Dobrakowska, who painted the cover, and Susan Everson, who composed it. Finally, thanks to Mats Heimdahl and Crystal King for their support from the University of Minnesota.

Obviously all errors are mine and I express my gratitude to the readers for their comments. Please send any questions, comments, or corrections to `the.footsteps.book@gmail.com`.

ABOUT THE AUTHOR

Krzysztof Frankowski (born 1932) is a founding member and professor emeritus of computer science at the University of Minnesota (USA). He became interested in mathematics in high school, more deeply so after winning an award at the first International Mathematical Olympiad in Warsaw. He got a bachelor's degree in mathematics at the University of Łódź, a master's degree in applied mathematics from the University of Warsaw, and a doctorate in philosophy (Ph.D.) from the Hebrew University in Jerusalem.

From 1958 to 1965 he worked as a researcher at the Weizmann Institute of Science in Israel, where he wrote programs on the WEIZAC, one of the world's first computers. From 1965 to 1997 he was a professor at the University of Minnesota in the Department of Computer Science. He retired in 1997 but continued to give guest lectures.

He is the husband of Elaine, the father of Michael and Daniel (both of them professionally deal with computers), is interested in music, and led the Polish choir Topola in Minneapolis for years. He often visits Poland and Israel.

www.ingramcontent.com/pod-product-compliance
Lightning Source LLC
Chambersburg PA
CBHW070436180526
45158CB00018B/1440